Quick Start to Data Analysis with SAS

Quick Start to Data Analysis with SAS

Frank DiIorio
ASG, Inc.

Kenneth A. Hardy
University of North Carolina

Duxbury Press
An Imprint of Wadsworth Publishing Company
I(T)P® An International Thomson Publishing Company

Belmont • Albany • Bonn • Boston • Cincinnati • Detroit • London • Madrid • Melbourne
Mexico City • New York • Paris • San Francisco • Singapore • Tokyo • Toronto • Washington

Statistics Editor: Stan Loll
Editorial Assistant: Claire Masson
Production Editor: Jerilyn Emori
Print Buyers: Barbara Britton, Diana Spence
Copy Editor: Judith Abrahms
Cover : Stuart Paterson
Printer: Quebecor Printing/Fairfield Inc.

Printed on acid-free recycled paper

Printed in the United States of America
1 2 3 4 5 6 7 8 9 10—01 00 99 98 97 96

For more information, contact Duxbury Press at Wadsworth Publishing Company:

Wadsworth Publishing Company
10 Davis Drive
Belmont, California 94002, USA

International Thomson Publishing Europe
Berkshire House 168-173
High Holborn
London, WC1V 7AA, England

Thomas Nelson Australia
102 Dodds Street
South Melbourne 3205
Victoria, Australia

Nelson Canada
1120 Birchmount Road
Scarborough, Ontario
Canada M1K 5G4

International Thomson Editores
Campos Eliseos 385, Piso 7
Col. Polanco
11560 México D.F. México

International Thomson Publishing GmbH
Königswinterer Strasse 418
53227 Bonn, Germany

International Thomson Publishing Asia
221 Henderson Road
#05-10 Henderson Building
Singapore 0315

International Thomson Publishing Japan
Hirakawacho Kyowa Building, 3F
2-2-1 Hirakawacho
Chiyoda-ku, Tokyo 102, Japan

Library of Congress Cataloging-in-Publication Data

DiIorio, Frank C.
Quick Start to Data Analysis with SAS / Frank DiIorio, Kenneth C. Hardy.
p. cm.
Includes index.
ISBN: 0-534-23760-6
1. SAS (Computer file) 2. Statisics—Data processing.
I. Hardy, Kenneth A. II. TItle.
QA276.4.D552 1995 95-18388
005.3—dc20

Contents

1 Introduction 1

SAS: The Complete Research Tool 1
Objectives 2
A Note About Syntax and Examples 2
    Syntax 2
    Examples 3
Organization 4
    Chapter by Chapter 4
What This Book Is *Not* 5

2 SAS Programming Concepts 7

Fundamentals 7
    Datasets 7
    Units of Work 8
    Syntax 9
    Defaults 10
    Running the Program 10
An Annotated Example 12
Recap 15

3 Preparing the Data for Analysis 17

Understanding the Input Data 18
    Missing Data 19
Naming the Dataset: The DATA Statement 19
Locating the Data: The INFILE, CARDS, and DATALINE Statements 20
Describing the Data Layout: The INPUT Statement 21
    List Input 21
    Column Input 22

Formatted Input *23*
INPUT Statement Formats *23*
Mixing Input Styles *24*
Which Style to Use? *25*
Common Problems *25*
Permanent Versus Temporary SAS Datasets 26
What's in the Dataset? *26*
Linking to the System: The LIBNAME Statement *27*
"Two-Level" Dataset Names *28*
Recap 28
Syntax Summary *28*
Content Review *29*

4 **Introduction to DATA Step Programming** **31**

Documenting Your Work: Using Comments 32
Reading SAS Datasets: SET 33
Syntax *33*
Examples *34*
Performing Calculations: Assignment Statements 35
Syntax and Usage Notes *35*
Examples *36*
Building Logical Expressions 37
Comparison Operators *37*
Logical Operators *38*
Examples *39*
Conditional Execution: IF-THEN-ELSE 39
Forms of IF-THEN-ELSE *40*
Examples *40*
A Note About Recoding *41*
Selecting Observations: OUTPUT 42
Increasing Variable Description: LABEL 43
Controlling Display of the Data: FORMAT 43
Recap 45
Syntax Summary *45*
Content Review *46*

5 **Combining Datasets** **47**

Methods 48
Concatenation *48*
Interleaving *48*
Matched Merge *48*
Terminology 48
Dataset Options 50

Concatenation 50
Examples 51
Interleaving 53
Example 53
Matched Merge 54
Syntax and Usage 54
Examples 56
Recap 59
Syntax Summary 59
Content Review 59

6 Introduction to Procedures 61

Statements Used with Most Procedures 62
Clarifying Content: TITLE and FOOTNOTE 62
Processing Data in Groups: BY 63
Controlling Appearance: FORMAT and LABEL Revisited 63
Selecting Observations: WHERE 64
Identifying Groups of Variables: Variable Lists 65
Displaying the Dataset: PRINT 66
Example 66
Identifying the Dataset 67
Identifying and Ordering Variables 68
Requesting Counts and Sums 69
Aesthetics 69
Example 70
Rearranging the Dataset: SORT 72
Specifying the Datasets 72
Specifying Observation Order: The BY Statement 72
Sorting Options 73
Getting Information about the Dataset: CONTENTS 74
Identifying the Datasets 74
Controlling the Amount of Output 74
Example 74
Recap 75
Syntax Summary 75
Content Review 76

7 A Statistical Roadmap for the SAS System 77

What's in a PROC Name? 77
A Warning 78
The Roadmap 78
Univariate Statistics 78
Simple Bivariate Tests and Measures of Association 79

Comparing Means and Analysis of Variance 79
Estimating Prediction Equations with Regression 80
Specialized Regression Models 81
Models for More than One Dependent Variable: Multivariate Statistics 81
Measurement and Scaling Techniques 82
Grouping and Predicting Group Membership 83
Navigating the Documentation Maze 83
Documentation 84
Recap 85

8 Statistics for Single Variables 87

Examining Distributions of Categorical Variables 87
Creating Frequency Distributions 87
Visualizing Frequency Distributions with Bar Charts 91
Examining Distributions of Continuous Variables 98
PROC MEANS: Basic Summary Statistics for Continuous Variables 102
Basic MEANS Syntax 102
Display and Analytical Options 102
Specifying Analysis Variables 103
PROC UNIVARIATE: Obtaining Detailed Statistics, Tables, and Graphs 104
Basic UNIVARIATE Syntax 104
Display and Analytical Options 107
Recap 109
Syntax Summary 109
Content Review 110

9 Statistics for Relationships with Continuous Dependent Variables 111

ANOVA Statistics for Categorical Independent Variables 112
Visualizing Group Distributions with PROC CHART 112
Visualizing Distributions with PROC UNIVARIATE 115
Testing Differences in Means for Two Independent Groups: PROC TTEST 116
Testing Differences in Means for Several Independent Groups: PROC GLM 118
Specifying Post Hoc Comparisons 121
Using PROC GLM for Two-Way to N-Way ANOVAs 124
Testing for Differences of Related Means 128
Regression: Statistics for Continuous Independent and Dependent Variables 133
Visualizing Bivariate Relationships Using PROC PLOT 133
Estimating Regression Models with PROC REG 136
Analysis of Covariance: Statistics for Both Categorical and Continuous Independent Variables 143

Using PROC GLM for ANCOVA 143
Interpreting PROC GLM ANCOVA Output 147
Using PROC REG for ANCOVA 148
Recap 152
Syntax Summary 152
Content Review 153
References 154

10 Statistics for Relationships with Categorical Dependent Variables 155

Examining Bivariate Relationships 156
Producing a Crosstabulation with Tests and Measures of Association 156
Entering Previously Crosstabulated Data 162
Testing for Differences Between Groups on a Rank-Ordered Dependent Variable 164
Logistic Regression: Estimating a Prediction Equation for a Dichotomous Dependent Variable 167
PROC LOGISTIC: Estimating Logistic Regression Models with Continuous Independent Variables 168
PROC CATMOD: Estimating Logistic Regression Models with Categorical Independent Variables 174
Recap 179
Syntax Summary 179
Content Review 180
References 180

11 More About SAS Programming 183

Options 184
Identifying System Settings 184
Frequently Used Options 185
DATA Step Programming Revisited 187
Grouping Related Statements: DO and END 187
Streamlining Calculations: Functions 187
Working with Dates 189
Background 190
Assigning Date Values 190
Manipulating Date Values 191
Displaying Date Values 192
Recap 193
Syntax Summary 193
Content Review 194

12 PROCs That Create Datasets 195

Aggregating by Groups: The MEANS Procedure 195
- *The PROC MEANS Statement 196*
- *Instructions for Aggregation: The CLASS Statement 197*
- *Specifying Analysis Variables: The VAR Statement 197*
- *Naming and Requesting Statistics: The OUTPUT Statement 198*
- *Examples 200*

Standardization of Data: The STANDARD Procedure 202
- *The PROC STANDARD Statement 202*
- *Specifying Analysis Variables: The VAR Statement 203*
- *Examples 203*

Rank-Ordering Data: The RANK Procedure 204
- *Background 204*
- *The PROC RANK Statement 204*
- *Specifying Input and Output Variables: The VAR and RANK Statements 205*
- *Examples 206*

Recap 207
- *Syntax Summary 207*
- *Content Review 208*

13 Recoding and Labeling with PROC FORMAT 209

Concepts 210
Syntax 210
Using Formats for Display 212
Analytical Use of Formats 214
Using Formats in DATA Steps 217
Recap 218
- *Syntax Summary 218*
- *Content Review 218*

14 Working with Character Data 221

Principles 221
- *Length 221*
- *Magnitude 221*

Specifying Length: The LENGTH Statement 222
Character-Handling Operators 223
- *The Concatenation Operator 223*
- *The Colon (:) Comparison Modifier 224*

Character-Handling Functions 224
- *Obtaining Variable Information: LENGTH 225*
- *Extracting Part of a String: SUBSTR 225*
- *Locating Strings Within a String: INDEX Variants and VERIFY 226*
- *Altering the Appearance of a String 227*

Recap 229
Syntax Summary 229
Content Review 229

15 **Putting It All Together 231**

Predicting Opinion About Taxes 231
Predicting Employee Turnover 242
Treating Fear of Snakes 255
Recap 259
Closing Comment 259
References 260

Appendix A **Using the Display Manager 261**

Display Manager Background 261
Syntax 265
A Sample Display Manager Session 266
Entering the Program 266
Running the Program 267
Moving Through the Output 268
Revising the Program 268
Saving Your Work 269
Leaving SAS 270
Retrieving Your Work 270
Customizing Your Environment with the KEYS Window 271
Syntax Summary 272
Command Line Commands 272
Line Commands 274

Appendix B **Resources 277**

Person-to-Person 277
Help Desks 277
"Power Users" 278
User Groups 278
Courses 278
Electronic Resources 279
Online Help 279
SAS/Assist 279
SAS Sample Library 279
The SAS-L List Server 280
World Wide Web 280

Hard Copy 281
Local Documentation 281
SAS Institute Publications 281
Books by Users 281

Appendix C Common Problems (and Solutions) 283

Divison by Zero 283
Unbalanced Quotes 284
DOing Without ENDing 285
Uninitialized Variables 286
Subtle Omissions of Periods 287
Display Manager "Stalls" 287
Data Type Conversion Messages 288
New Character Variables Appear Truncated 289
Subsetting into 0 Observations 289
Complaints About Correct Syntax 290
Unbalanced Parentheses 291
"Dataset Not Found" Messages 292
Every Calculated Value Is Missing 292
Conflicting Data Types 293
All Calculations Are Missing 294

Index 295

Preface

Quick Start to Data Analysis with SAS is an introduction to the SAS System for students in beginning- to intermediate-level statistics courses. SAS software has developed a well-earned reputation for the high quality of its analytical tools and, on the flip side, the sometimes baffling syntax required for their use. Quick Start distills the voluminous SAS Institute documentation into a brief but comprehensive introduction to the SAS System.

We present a small fraction of SAS's capabilities. But it is, given our collective forty-four years' experience with the product, the fraction most relevant for those taking their first steps into SAS software for data analysis. On completion of this book you should have a clear understanding of the basics of the SAS System and be able to perform a variety of data management, reporting, and analysis tasks. You may even develop an interest in learning more about the SAS System. If we simply help you get through your homework and research more effectively and enjoyably, though, we've done our job!

Level of Coverage

By any measure, the SAS System is huge. It has a complete set of facilities for data management, reporting, and user interface design. The breadth of its analytical capabilities is unmatched in the marketplace—students of econometrics, operations research, and virtually any statistical technique will find the tool they need here.

The downside of this scope and power is the SAS System's rather steep learning curve. A beginner just doesn't start a SAS session, snoop around through the menus, and figure things out. A little time is required to get a feel for how SAS programs are constructed and other aspects of what makes SAS tick.

Our goal in this book is to get you to the point of being able to use the SAS System as quickly as possible. This speed is always tempered with our desire to help you avoid the beginners' errors and misconceptions we've noticed over the years. The only feasible way to maintain this balance between speed and safety is to carefully choose a small subset of the SAS language and explain it in reasonable depth. We also defer the task of teaching statistics to *bona fide* statistics textbooks and courses. Finally, as you read the text, remember that if you think SAS should be able to do more in a given area, it probably can (we give you resources for further study in Appendix B and at the end of many chapters).

Text Organization

The text is divided into fifteen chapters and three appendices. Chapter 1 introduces our syntax conventions and provides a detailed, chapter-by-chapter description of the book. Chapter 2 presents the basic concepts of SAS programming. Chapter 3 puts some of these concepts to work, showing how to read data and prepare it for analysis.

Chapter 4 shows techniques for calculations, conditionally performing actions, and enhancing the display of data. In Chapter 5 we give some techniques for combining data from multiple sources. We begin to use the data for analysis and display in Chapter 6, which describes the use of several popular SAS procedures.

As we noted in the preceding section, the SAS System's analytical capabilities are vast. Finding just the right tool in such a large toolbox can be frustrating. Chapter 7 attempts to alleviate some of the frustration by discussing tool selection and providing a road map of SAS procedures.

Chapters 8 through 10 walk you through use of single-variable, bivariate, and multivariate statistics. We introduce the procedures gradually, give many examples, and summarize the syntax at the end of each chapter.

Chapters 11 through 14 may be read independently. They address techniques for creating summary and transformed datasets, more advanced programming techniques, and working with special types of data. In Chapter 15 we present three mini-case studies, which tie together many of the concepts presented in the preceding chapters.

The majority of the book is devoted to describing *what* SAS can do. Appendix A takes a different approach and discusses *how* you actually run SAS. The Display Manager interface is discussed in detail. Appendix B gives an overview of how to learn more about the SAS System. Various print, electronic, and other resources are enumerated. Lastly, in Appendix C we present a series of common "gotchas" encountered by beginners (and, sorry to say, sometimes even by SAS professionals). The errors and one or more solutions are presented.

Text Features

As noted earlier, we trade off breadth of coverage for depth and pace of explanation. We present new features and capabilities gradually, using annotated examples. A syntax summary is found in the last section of each chapter. This recap is used as a quick reference once you become familiar with the chapter's tools. A variety of good programming practices is discussed in the text and put into practice in the examples. It is our hope that you will adopt some of these conventions in your programming and realize the benefits of writing clear, well-documented code.

Using This Book

We anticipate that this book will be used primarily by students either taking statistics courses or conducting their own research. Everyone should read Chapters 1 and 2, since these provide important background and conceptual details. If your research or class data are provided as SAS datasets, you can skip Chapter 3's sections on reading raw data. Read the last pages of Chapter 3, "Permanent Versus Temporary SAS Datasets," since this is the key to being able to access your class or research SAS datasets. If your data are ready to analyze and don't require any calculations or transformations, skip to Chapter 6, which presents useful background information about SAS procedures. Otherwise, read Chapter 4 and, if you need to combine data from multiple sources, Chapter 5. Your instructor may direct you to a particular analytical procedure. If this is not the case and you need some guidance about selecting the appropriate tool, read Chapter 7. The next chapters, 8 through 10, describe usage of procedures for everything from simple frequency distributions to logistic regression. They are reasonably independent of each other.

Chapters 11 through 14 are independent, stand-alone treatments of special topics in SAS programming. Chapter 13 may be of particular interest. It discusses SAS formats, powerful tools to add descriptive labels to values as well as recode and collapse ranges of variables.

If you want to see a series of complete SAS programs and their output, scan Chapter 15. It contains three mini-case studies and gives you an idea of what realistic "real-world" research looks like in the SAS System.

The appendices do not have to be read in order. Appendix A is a helpful reference for interactive users running the SAS Display Manager. If you are running in an interactive (non-batch) environment, it may be worthwhile to read or skim this appendix before going too far in the text. Appendix B points you to resources for learning more about the SAS System. Finally, Appendix C presents common problems encountered by beginners. Periodically review it to stay aware of some of the more popular "gotchas."

Acknowledgments

We owe a debt of thanks to many people for their assistance and encouragement. We had the good fortune to have early versions of this work reviewed by the following eagle-eyed, insightful, and sometimes inflammatory people: Gordon A. Allen, Miami University; Michael J. Camasso, Rutgers University; Judith Ekstrand, San Francisco State University; Rober Hamer, Rutgers University; Andrew H. Karp, Sierra Information Services; Howard Nusbaum, University of Chicago; and Craig Stewart, Indiana University. We listened to all of their criticisms and took most of them to heart. The product you are reading now is, we are sure, much better for their comments.

The staff of Wadsworth Publishing Company provided invaluable editorial and production assistance and kept us on track by being adept with both carrot and stick. They are: Stan Loll, Statistics Editor; Jerilyn Emori, Senior Production Editor; and Stephen Rapley, Art Director. Special thanks to Judith Abrahms for her meticulous copyediting efforts and to Cathy Jarcho for her assistance converting our rough pages into the polished form you are now reading.

Personally, from Frank: After the long, arduous birth of my first book, I vowed "never again," and my family breathed a sigh of relief. Obviously, I broke that promise, yet my wife, Elizabeth, and daughters, Amelia and Sophia, are still talking to me. As a small way of making up the lost hours of family time, I would like to dedicate this book to them. This time, I'll only vow to stop making promises I can't keep!

Personally, from Ken: Thanks to the Institute for Research in Social Science at the University of North Carolina at Chapel Hill for giving me some uninterrupted time to work on this book. Looks like I'll have to start showing up again on Fridays.

I also wish to thank all the graduate students and faculty members who have appeared at my door over the last twenty-three years with questions about using SAS for data management and statistical analysis. A lot of what I know about SAS I learned in response to their questions. I owe a considerable debt to those who have served as consultants with the institute's statistical and computer consulting service. They have been my tutors as well as my students. Special thanks go to Chris Agnew, who took the time to read and comment on many chapters even while he was very busy running the consulting service and trying to land a "real" job.

I can't express enough appreciation for my wife, Nancy, and son, Travis, who had the good sense and patience to leave me alone when I was preoccupied with writing. Their love and support helped me get the job done.

Introduction

1

The calculation of statistics doesn't have to hurt any more. Years ago, before the arrival of computers and "user-friendly" interfaces, statistical analyses were tedious affairs requiring hours of hand calculations. Today the situation is vastly different. Dozens of statistical software packages are available, running in every computing environment. Among the market leaders are SPSS, SYSTAT, STATA, S, Minitab, and the SAS System. All are adequate, all are suitable for a wide range of analyses, and most are easy to use.

Consider, though, the scope of the research task. Analyzing data assumes there are data ready to analyze! Tasks such as extracting analytical data from large databases, cleaning up errors, recoding variables, and documenting the contents of the datasets must be performed before any analysis takes place. This often takes as much time as the actual analysis, if not more. What is really needed is a tool that *manages* data as well as *analyzes* it. One can successfully argue that SAS software addresses these two needs better than any product currently available. This book introduces you to using the SAS System for statistical analysis.

SAS: The Complete Research Tool

Many competing products have advantages over the SAS System. Some of these advantages are significant: SPSS for Windows makes data entry simple and intuitive; Minitab's syntax is simple and powerful; S+ enables the researcher to create leading-edge analyses. All lack the powerful data management capabilities found in the SAS System (referred to throughout the book simply as SAS). SAS not only lets you prepare and maintain data but also places a wide variety of analysis and reporting tools at your disposal. Within the boundaries of SAS software you can read data, clean it, update it, and analyze and report it.

These advantages are not lost on organizations in the "real world." SAS software penetration in private industry and government is significant. Nearly every Fortune 1000 company uses SAS products, as do a wide range of government and research organizations. This means that the tool you use for classroom work will provide you with a marketable skill in the job market — there is a "carryover" into real-life applications.

Objectives

This book presents an introduction to SAS as a data management and analysis tool. It gives the reader a solid introduction to the capabilities of SAS as both a data management and a statistical tool. When you've finished the book, you will be able to

- Read data in a variety of formats
- Transform the data as needed to perform different statistical analyses
- Perform a wide range of basic statistical analyses used in the health, management, behavioral, and social sciences
- Understand how the SAS System "thinks," which will give you a good grounding for further, more advanced work with the software

The book does *not* attempt to transform you into a skilled programmer. Nor does it succumb to the urge to show off the "neat tricks" that are possible with SAS. It simply attempts to give you the tools necessary to perform data analysis, encouraging you along the way to learn more about this extremely powerful and effective software.

A Note About Syntax and Examples

Syntax

There are many ways to explain the syntax of the SAS language. Throughout this book, we use the conventions employed by SAS Institute in its documentation. This has a logistical advantage: If you need to use the Institute's documentation for techniques not described in this book, the transition to the voluminous Institute literature will be relatively easy. The conventions are summarized below.

- UPPERCASE indicates a keyword. It does not have to be entered in uppercase in your program, but must be spelled as shown.
- *Italic lowercase* indicates an item that you name. These include, but are not limited to, dataset and variable names (described in the next chapter).
- ... (ellipsis) indicates items that may be repeated as necessary. A common example of this is a list of variables to be included in an analysis.
- | (vertical bar) indicates that only one of several options or keywords may be chosen. The list is usually enclosed in angle brackets (the characters "<" and ">").

These elements are demonstrated in the syntax descriptions for two common routines in the SAS language, shown below. PROC FREQ produces 1- to *n*-way frequency distributions. PROC CHART produces bar charts. Examine these descriptions to see how the syntax conventions are used. Do not be concerned about the particulars of the PROCs (or, for that matter, about what a PROC is). These will be covered in later chapters.

```
PROC FREQ <DATA=dataset_name> <ORDER=FREQ>;
    TABLES varname1 <varname2 ...>
    </ <MISSING> <NOCUM> >;

PROC CHART <DATA=dataset_name>;
    HBAR|VBAR varname1 <varname2 ...> </
    <ASCENDING|DESCENDING>
    <DISCRETE> <LEVELS=nbars>
    <<MIDPOINTS = val1 val2 <val3 ... >|
    <MIDPOINTS = low TO high <BY step>>
    <MISSING> <NOSTATS> <SPACE=nspace>>;
```

Examples

We include many examples of SAS output throughout the book. Two cautionary notes need to be made. First, the programs that created the output usually used two SAS *system options* (described in Chapter 11) to display output left-justified on the page and to use only 78 print positions. Second, most of the programs were written on and run on Versions 6.08 and 6.10 of the SAS System running under Microsoft Windows, Version 3.11.

Why is this relevant? Because the output you generate using your version of SAS may not appear exactly the same as the output shown in this book. The standard, default settings for the system options may be different, and you may not be running SAS in the Windows environment. What is important to remember is that even though the *exact* appearance may differ, the basic *content* of the output is the same regardless of pagination, centering, print positions, and operating system.

Organization

Throughout most of the book, we attempt to place data management and statistical concepts ahead of computing terminology. That is, we structure the discussion along the lines of the research process rather than approaching SAS as a programming language. Rather than discuss the FREQ, TTEST, and CORR procedures outside of their analytical context, for example, we "bundle" them into one chapter, "Statistics for Describing Relationships Between Two Variables." The *content* in either method of presentation would be the same. The difference lies only in the way it is presented; using the statistical context as the basis for each chapter makes it easier for the reader to find the appropriate analytical technique.

We supplement the discussion of the SAS language by using a few simple pedagogical techniques. Short examples are used to clarify nuances of syntax. The material is presented in tabular form as often as possible. Examples are liberally annotated and build on previous examples, gradually becoming more complex. We note common problems faced by beginners. Also, to be fair, we suggest possible resolutions of these problems. A syntax summary and a content review conclude each chapter.

Chapter by Chapter

Chapter 2 introduces the fundamentals of the SAS language. The *SAS dataset* is introduced, and we show how SAS programs are structured. Several techniques for running a SAS program are outlined, as are the basics of the language's syntax. We present a sample program and go over its output in detail.

We begin to read data and store SAS datasets in Chapter 3. Only a few tools are needed for this, and all are discussed here. Chapter 4 presents basic features used to perform calculations, conditionally execute program statements, and control what is written to the SAS dataset. Facilities for improving the readability of the program's output are also discussed.

The research process often requires the combining of data from multiple sources. The SAS System provides a simple yet powerful set of tools for combining datasets. These are covered in Chapter 5. Some important utilities are presented in Chapter 6, which also discusses features of the SAS language that may be used with any of the analytical procedures discussed later in the book.

In Chapter 7, we present a statistical "roadmap" for the SAS System. The roadmap classifies statistical methods (univariate, bivariate, and multivariate) and identifies the SAS procedures typically used with these techniques. Chapters 8 through 10 demonstrate how these procedures are used. Simple applications are always presented first, followed by more advanced examples.

Chapter 11 addresses some intermediate SAS programming topics. We show you how to adjust SAS System option settings, use computational shortcuts, and handle date-oriented data. Chapter 12 discusses procedures that create SAS datasets. Chapter 13 presents a powerful procedure you can use to recode and group data. In Chapter 14, we outline usage of character, or text-oriented, data. A variety of tools are presented for reading, manipulating, and displaying this type of data.

Chapter 15 presents three mini case studies. These examples tie together many of the concepts presented in earlier chapters. In doing so, they give you an idea of how to use SAS to address common research problems. Three appendices supplement the book. Appendix A explains how to use the interactive interface to the SAS System, the Display Manager. Appendix B identifies a wide variety of resources (some of them free) that you can use to supplement this book. Finally, Appendix C discusses a series of common problems encountered by beginners.

What This Book Is *Not*

We have a lot of ground to cover in a fairly small space. It will be helpful to note here what is not covered in the book. First, the book describes a fairly small subset of the SAS programming language. It is *not* a complete reference work presenting the basic elements of the language. These are covered in the copious documentation available from SAS Institute (see Appendix B for details). It does, however, cover a highly useful subset of the language in enough detail for you to safely, and perhaps even enjoyably, perform basic analytical tasks.

The book is not intended to be a statistics text. Such instruction should be available from the main textbook in the course you are taking. Trying to cover both statistical techniques and the SAS language would render any authors (and readers!) weary. We elected not to try it! We *do* intend to present you with a clear, well-paced introduction to the SAS programming language, to whet your appetite for further study, and to show you that statistical analysis with SAS System software can be a pleasant experience.

SAS Programming Concepts

2

Rather than immediately jump into the presentation of SAS statements, we will focus for a few pages on some of the principles that underlie all the programming you will be doing in this book. Readers familiar with programming languages and databases will see some familiar terminology and concepts. Those of you who are new to computer usage will see that much of what is presented is reasonably straightforward: the way SAS "thinks" and the way it views data are simple and appealing. To reiterate the opening of Chapter 1, it doesn't have to hurt to become productive with SAS software.

We first look at the components of a SAS program: datasets and tools for reading, modifying, and analyzing data are presented. We then review methods for running the program. Finally, we put the pieces together in an annotated sample program. When reading the chapter, pay more attention to the overall "feel" of the program than to syntax particulars: *what* is being done is, for the time being, more important than *how* the work is accomplished.

Fundamentals

Datasets

Using SAS to analyze data evolves around the creation and use of SAS datasets. These are specially laid-out versions of your data that allow the SAS System to read and use data in a wide range of statistical, graphical, and reporting routines. SAS programs read "raw" data, create SAS datasets, perform transformations, and perform analyses. The raw data are usually a collection of data points stored in a computer file that can be thought of as a table of rows and columns. The *rows* represent the unit of observation: a respondent to a questionnaire, a company, a geographic area, and so on. SAS refers to rows as *observations.*

Each *column* of raw data represents a measurement or characteristic of the observation. These could be items such as the response to a question in a survey, the number of company employees in 1990, or the population density of a county. SAS refers to the table's columns as *variables.* They are either character (for instance, names and job descriptions) or numeric (for instance, salaries, ages, and test scores). The character or numeric characteristic is often referred to as a variable's *data type*.

The data type of a variable is a *programming* construct. It does not always have implications for a variable's level of measurement, which is a *statistical* concept. Ratio and interval scale variables will necessarily be numeric. Ordinal and nominal scales, however, may be represented by variables of either data type. Examples throughout this book use both character and numeric SAS variables for ordinal and nominally scaled variables.

All the data for a study do not necessarily have to be located in a single SAS dataset. If it makes more sense to enter data for, say, one year of a survey in one file and those for another year in a separate file, you can do so. Chapter 5 presents several methods for combining SAS datasets.

It is desirable, but not required, that data be complete. That is, every observation should have legitimate values for every variable. The "real world" is seldom so accommodating: for example, people may refuse to answer questions on a survey, or a test may not have been administered at the time required by a clinical trials protocol. SAS has rules for handling instances of *missing values* for variables. This important topic is dealt with throughout this text. For now, just remember that missing values are a normal consequence of the research process and that SAS's default, or automatic, reaction to them is usually appropriate.

Observations and variables are the main components of the SAS dataset. The SAS dataset you use for statistical analysis can be constructed not only from raw data but from other SAS datasets as well. All the statistical and reporting procedures in this book require data to be stored as SAS datasets. Fortunately, only a few statements are needed to convert raw data into SAS datasets or to combine SAS datasets.

Units of Work

Any SAS program's activity can be carried out by two general units of work. The *DATA step* reads SAS datasets or raw data, performs transformations, creates new variables, and recodes existing variables. It is the primary mechanism for creation of SAS datasets.

The second unit of work is the *procedure*, usually referred to simply as a PROC. The PROC analyzes and reports on the data. The specifics of its use are not important now. Just remember that there is no prescribed format for a program; there can be as many DATA steps and PROCs in a program as you need. The only restriction is that the order must make logical sense. You shouldn't, for example, try to run a scatterplot on a dataset before creating it!

The building blocks of both DATA steps and PROCs are *statements*. One or more statements make up the DATA step or PROC. A statement is usually identified by a keyword that implies its function: the INPUT statement, for example, reads (inputs) a record from a raw dataset, the FREQ procedure produces frequency distributions, and the RUN statement signals the end of a unit of work and tells SAS to *run* the DATA step or PROC just completed.

Syntax

How are statements put together? This varies, of course, according to the task at hand: calculations require different phrasing from that used for bar charts, for example. A few features are common to all statements:

- ❑ Names you assign to variables and datasets can be no longer than eight characters, must begin with a letter or an underscore (_), and can include only letters, numbers, and underscores.

- ❑ In most situations, the case of the statement is not important. Some programmers prefer EVERYTHING IN UPPERCASE; others consistently use lowercase. It is a matter of taste.

- ❑ Statements can extend across more than one line. It is often preferable to write programs this way, since the extra blank space improves program legibility.

- ❑ Every statement ends in a semicolon (;).

The preceding comments about taste and legibility are important. A readable program is much easier to come back to after a long absence than is a program that was hastily thrown together. As you continue to use SAS software you'll develop a style and consistency of writing that improve the readability and accuracy of your programs.

Defaults

The syntax described above gives you fairly generous boundaries within which you can wander, but there's still the question of how many statements are actually needed to carry out a basic research task. Happily, the answer is often "Not many." This is because SAS software was designed with intelligent *defaults*, or assumptions about how people typically go about their work. There are defaults for every aspect of the SAS program: procedure results centered on the page, disk storage reserved for variables, analytical methods, handling of incomplete data in procedures, and many other features. Later on, in Chapter 11, we will see how to take control of the SAS environment and override these defaults. For now, though, just keep in mind that most SAS programs are fairly short because the SAS System's defaults are working in your favor. This is one of SAS's greatest strengths.

Running the Program

Of course, identifying the data source and the required DATA steps and PROCs is one thing. Running the program and retrieving its output is quite another. Unfortunately, the commands for doing these things vary greatly from school to school and from company to company. Add to this SAS's ability to run on different hardware and operating-system environments, as well as in different styles, or modes, and the situation becomes even more complex.

The specifics of running SAS software at your site should be dealt with by a course instructor or by user-service, "help desk" personnel. What we can discuss here in general terms are your choices as to how you'll run SAS and the output you'll get when running a SAS program.

Modes of Execution You usually have several different ways of running SAS at your disposal. Common modes of program execution are *batch* and *interactive*. Your choices may be limited by your installation's type of computer and by campus or company policies.

In *batch* processing, you write a program with a text editor, then submit it for processing by the operating system. The program runs when resources are available; the results are sent to a printer and/or can be viewed at your terminal. Batch mode is typically used when large volumes of data are being processed and the program is error-free. One of its benefits is that since it runs "in the background," you are free to perform other tasks while it executes. The drawback is that syntax and logic errors cannot be detected and remedied until the program terminates. Immediacy is lost and program development speed is slowed.

Interactive processing, in contrast, displays results as soon as you send, or submit, statements to SAS. This is most often done by interacting with the SAS

Display Manager, a full-screen environment that allows you to write and execute programs and view results (Display Manager is discussed at greater length in Appendix A). You can see messages and output immediately, and make corrections and enhancements as needed. Interactive mode is usually used for development of programs and for analysis and handling of small datasets.

Program Output SAS programs usually produce two types of output: the *SAS Log*, usually referred to simply as the Log, and program *output*. The Log is produced whether the program runs successfully or not. The Log can be an extremely valuable resource! It contains the program statements you wrote and SAS's reactions to them. This includes such items as storage and memory resources used, notes about the source of the data and the size (expressed as observations and variables) of the SAS dataset created, warnings about problems with calculations, and errors such as invalid syntax, references to nonexistent variables, division by zero, and so on. Before looking at the output listing, look at the Log to make sure there are no messages about problems in carrying out the analysis.

Exhibit 2.1 illustrates how the SAS System reacted to two computational problems. The first is in line 4, where we divided by a missing value. The second is in line 7, where we divided by zero. The SAS System displays a note or an error message and gives other useful diagnostic information.

Exhibit 2.1

Log Showing Computational Errors

```
1    data zero_div;
2    num1   = 5;
3    denom1 = .;
4    calc1  = num1 / denom1;
5    num2   = 3;
6    denom2 = 0;
7    calc2  = num2 / denom2;
8    run;

NOTE: Division by zero detected at line 7 column 15.
NUM1=5 DENOM1=. CALC1=. NUM2=3 DENOM2=0 CALC2=. _ERROR_=1 _N_=1
NOTE: Missing values were generated as a result of performing an
      operation on missing values.
      Each place is given by: (Number of times) at (Line):(Column).
      1 at 4:15
NOTE: Mathematical operations could not be performed at the following
      places. The results of the operations have been set to missing
      values.
      Each place is given by: (Number of times) at (Line):(Column).
      1 at 7:15

NOTE: The data set WORK.ZERO_DIV has 1 observations and 6 variables.
```

The program output is usually created by PROCs. It takes many forms, among them statistical tables and matrices, charts, maps, and simulated three-dimensional renderings of data. As a rule of thumb, think of the Log as an audit trail and the output as the analytical result of the program.

An Annotated Example

The concepts and terminology discussed above are best illustrated by a simple, realistic program. This section presents the program, the data it uses, and the SAS Log and listing files created by the program. Most of the comments are enclosed in boxes within the exhibits themselves; an arrow connects each box with the item(s) described.

Exhibit 2.2 shows a partial listing of a sample raw dataset. It consists of data from the U.S. Federal Aviation Administration (FAA) about major airports in the United States. The columns, or variables, are airport name, city, state, and various measures of size (number of passengers, tons of freight, and so on). The rows, or observations, are the individual airports. The exhibit lists the first 10 observations in the raw dataset. Notice that the variables are arranged so that in every observation the same columns contain the same types of data — columns 44 and 45 always contain the state code, for example. The data do not always have to be this tidy — methods for reading this and more problematic, free-form data are presented in Chapter 3.

Exhibit 2.2

Partial Listing of FAA Dataset

Observations / rows

Column ruler

```
----+----1----+----2----+----3----+----4----+----5----+----6----+----7----+----8----+--

HARTSFIELD INTL      ATLANTA                 GA 285693 288803 22665665 165668.76  93039.48
BALTO/WASH INTL      BALTIMORE               MD  73300  74048  4420425  18041.52  19722.93
LOGAN INTL           BOSTON                  MA 114153 115524  9549585 127815.09  29785.72
DOUGLAS MUNI         CHARLOTTE               NC 120210 121798  7076954  36242.84  15399.46
MIDWAY               CHICAGO                 IL  64465  66389  3547040   4494.78   4485.58
O'HARE INTL          CHICAGO                 IL 322430 332338 25636383 300463.80 140359.38
DALLAS/FT WORTH INTL DALLAS/FT WORTH         TX 266737 269665 22899267 142660.95  86706.76
LOVE FIELD           DALLAS/FT WORTH         TX  39481  40196  2882836   2216.70    242.87
STAPLETON INTL       DENVER                  CO 154067 156293 11961839  67345.75  38043.73
DETROIT CITY         DETROIT                 MI   6828   7162   362655    258.08      0.00
```

Variables / columns

A program to read the data and perform some analysis of it is shown in Exhibit 2.3. The mechanics and syntax of the program are, at this point in the book, not relevant. What *is* important is to understand the flow of execution: a dataset was read, then analyzed. That is, a DATA step had to precede the PROCs that used it. A DATA step creates SAS dataset "airport", then a PROC lists the data (PRINT procedure), followed by a PROC that generates some simple univariate statistics (MEANS procedure) and a PROC that produces frequency tables (FREQ procedure).

Exhibit 2.3

Sample Program to Read and Analyze Raw Data

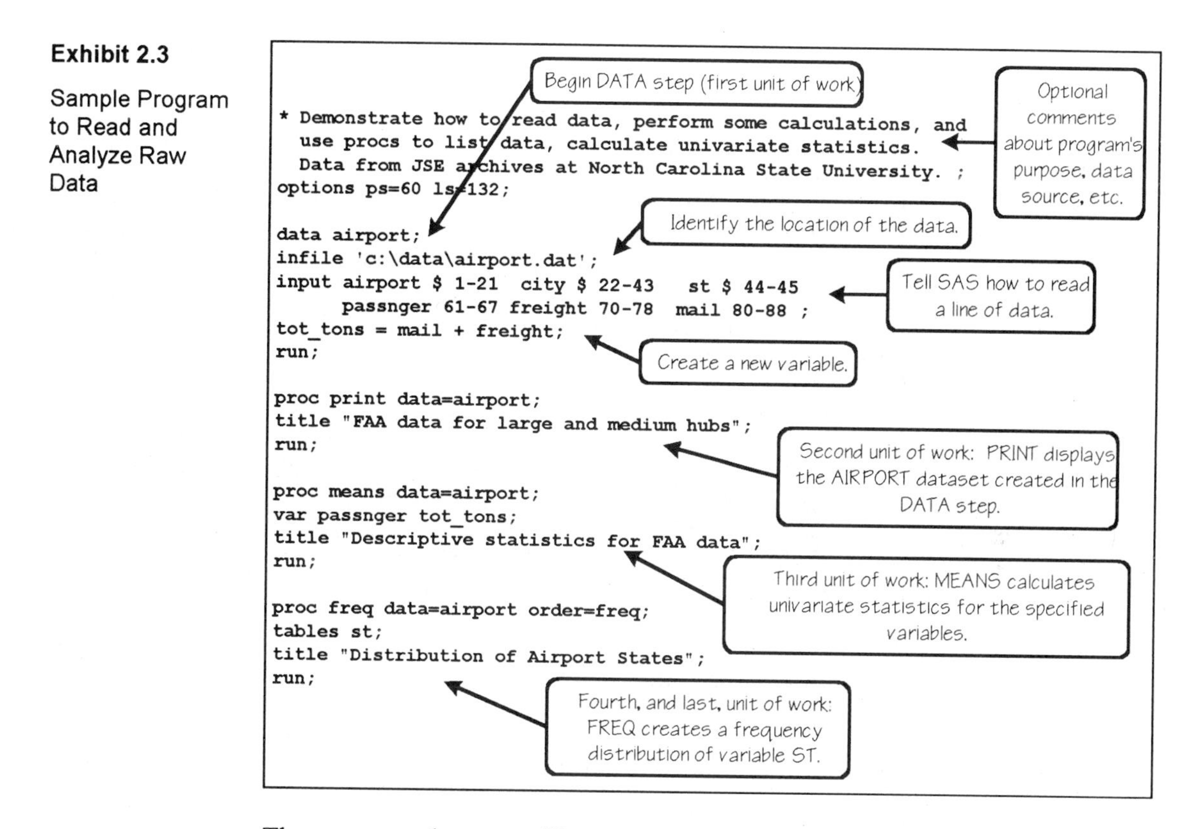

```
* Demonstrate how to read data, perform some calculations, and
  use procs to list data, calculate univariate statistics.
  Data from JSE archives at North Carolina State University. ;
options ps=60 ls=132;

data airport;
infile 'c:\data\airport.dat';
input airport $ 1-21  city $ 22-43   st $ 44-45
      passnger 61-67 freight 70-78  mail 80-88 ;
tot_tons = mail + freight;
run;

proc print data=airport;
title "FAA data for large and medium hubs";
run;

proc means data=airport;
var passnger tot_tons;
title "Descriptive statistics for FAA data";
run;

proc freq data=airport order=freq;
tables st;
title "Distribution of Airport States";
run;
```

Begin DATA step (first unit of work)
Optional comments about program's purpose, data source, etc.
Identify the location of the data.
Tell SAS how to read a line of data.
Create a new variable.
Second unit of work: PRINT displays the AIRPORT dataset created in the DATA step.
Third unit of work: MEANS calculates univariate statistics for the specified variables.
Fourth, and last, unit of work: FREQ creates a frequency distribution of variable ST.

The program, when run, will generate a SAS Log similar to the one shown in Exhibit 2.4. The Log resembles the program in Exhibit 2.3, but with a few additions. Remember, it is SAS's reaction to your program. Therefore, it will display the program along with items such as the size of the dataset(s) created, the size of the raw datasets read, how long it took to complete each unit of work, and so on. We see that the SAS System's reaction to the program was favorable. Only NOTEs were generated. If we had severe data problems, tried to perform an illegal calculation, or had a syntax error, SAS would display an ERROR or a WARNING in the Log and execution of the program would halt. Examine Exhibit 2.4 more for "look and feel" than for particulars about syntax and logic.

Exhibit 2.4

SAS Log for Sample Program

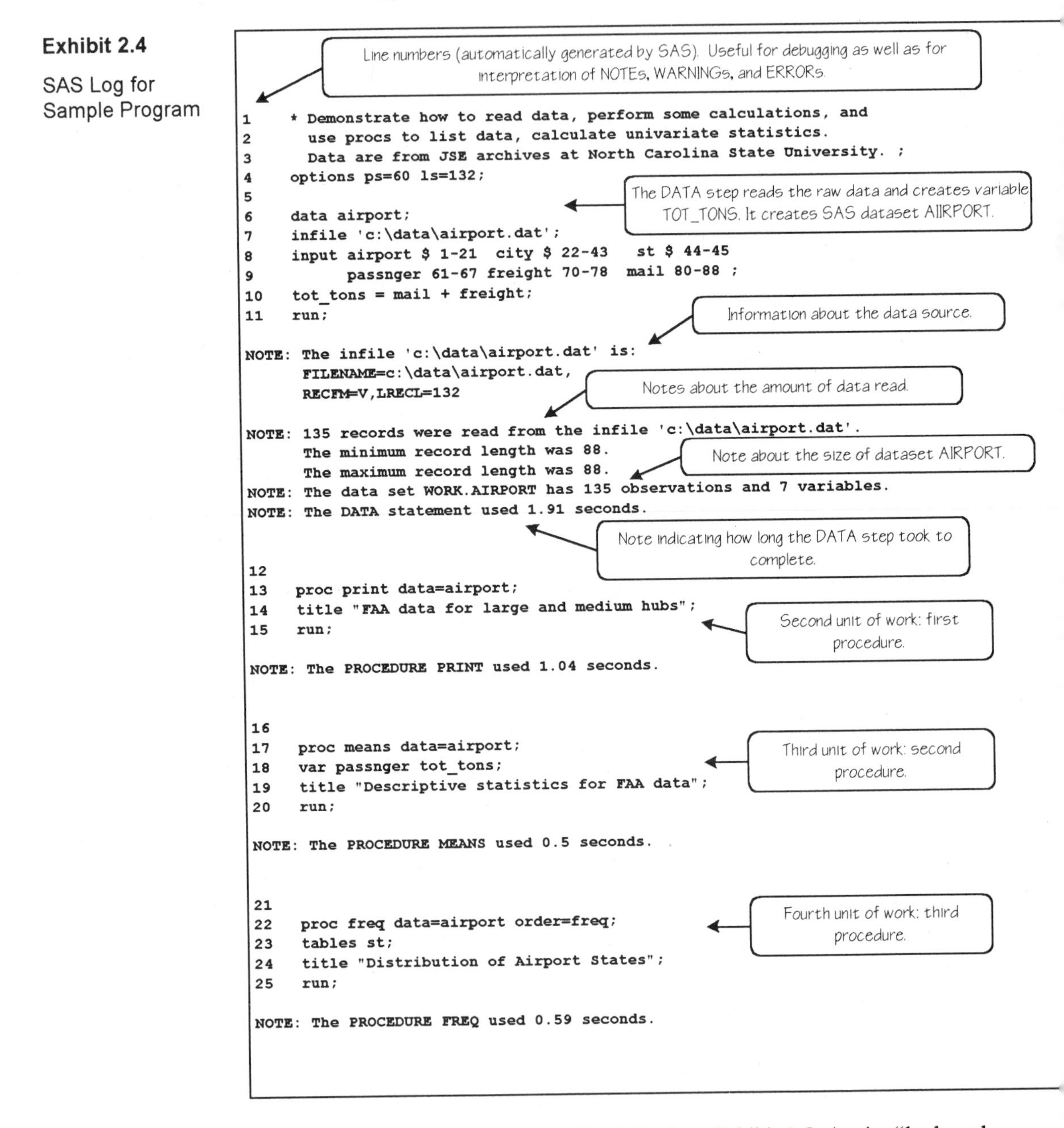

```
1     * Demonstrate how to read data, perform some calculations, and
2       use procs to list data, calculate univariate statistics.
3       Data are from JSE archives at North Carolina State University. ;
4     options ps=60 ls=132;
5
6     data airport;
7     infile 'c:\data\airport.dat';
8     input airport $ 1-21  city $ 22-43   st $ 44-45
9           passnger 61-67 freight 70-78  mail 80-88 ;
10    tot_tons = mail + freight;
11    run;

NOTE: The infile 'c:\data\airport.dat' is:
      FILENAME=c:\data\airport.dat,
      RECFM=V,LRECL=132

NOTE: 135 records were read from the infile 'c:\data\airport.dat'.
      The minimum record length was 88.
      The maximum record length was 88.
NOTE: The data set WORK.AIRPORT has 135 observations and 7 variables.
NOTE: The DATA statement used 1.91 seconds.

12
13    proc print data=airport;
14    title "FAA data for large and medium hubs";
15    run;

NOTE: The PROCEDURE PRINT used 1.04 seconds.

16
17    proc means data=airport;
18    var passnger tot_tons;
19    title "Descriptive statistics for FAA data";
20    run;

NOTE: The PROCEDURE MEANS used 0.5 seconds.

21
22    proc freq data=airport order=freq;
23    tables st;
24    title "Distribution of Airport States";
25    run;

NOTE: The PROCEDURE FREQ used 0.59 seconds.
```

Line numbers (automatically generated by SAS). Useful for debugging as well as for interpretation of NOTEs, WARNINGs, and ERRORs.
The DATA step reads the raw data and creates variable TOT_TONS. It creates SAS dataset AIIRPORT.
Information about the data source.
Notes about the amount of data read.
Note about the size of dataset AIRPORT.
Note indicating how long the DATA step took to complete.
Second unit of work: first procedure.
Third unit of work: second procedure.
Fourth unit of work: third procedure.

So what do we get for our coding effort? Look at Exhibit 2.5. Again, "look and feel" are more important than data values and statistics in the listing. The exhibit reveals some characteristics of the way SAS processes procedures. First, unless the user overrides the SAS System, the page number and the date and time the program ran are displayed at the top of each page. Second, the output from each procedure begins on a new page. Third, we see that it is possible to perform simple tasks with very little coding effort. Readers familiar with C, FORTRAN, or other languages can appreciate the amount of work done by the few statements in the program.

Exhibit 2.5

Output Listing for Sample Program

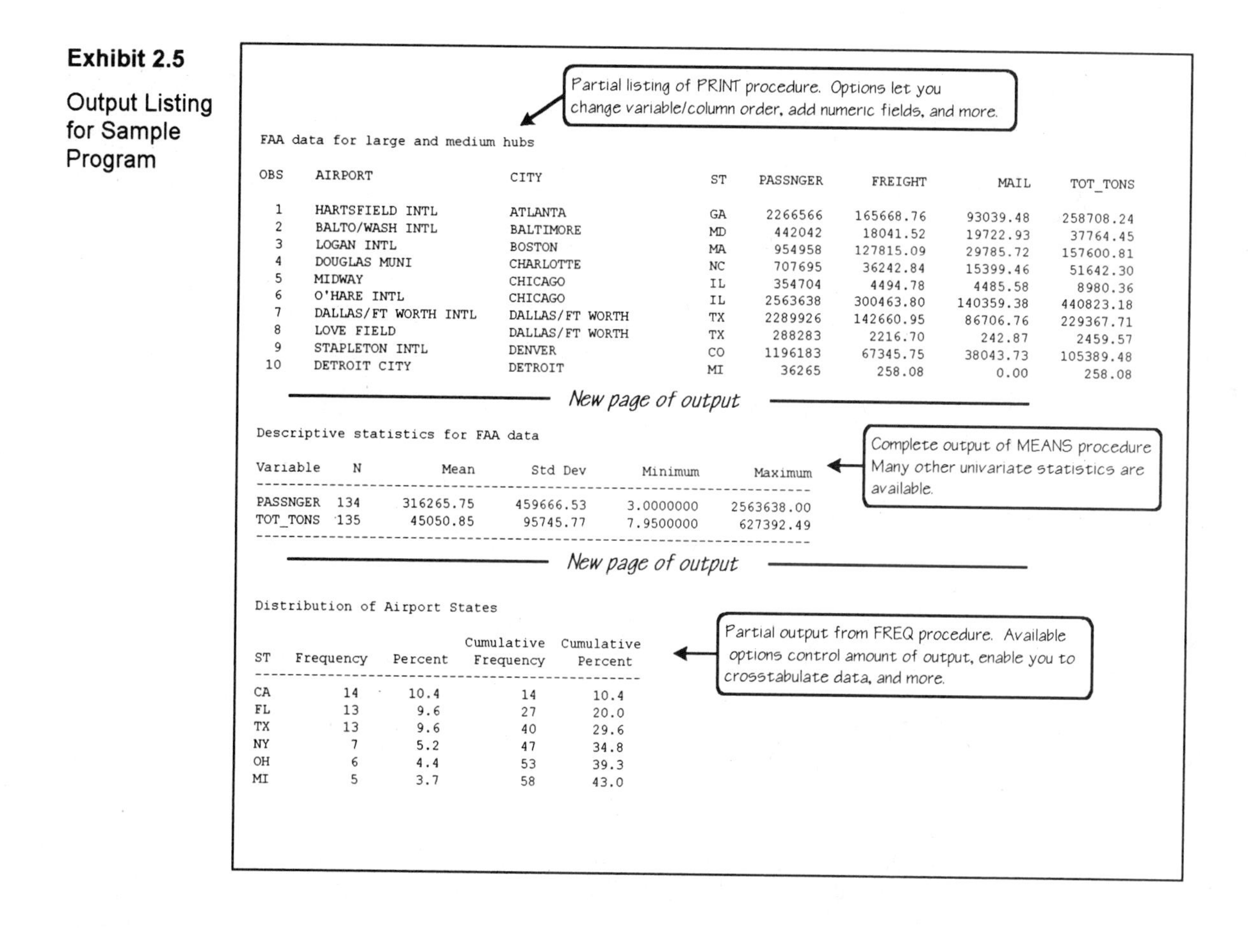


```
FAA data for large and medium hubs

OBS  AIRPORT                CITY               ST  PASSNGER   FREIGHT       MAIL     TOT_TONS

  1  HARTSFIELD INTL        ATLANTA            GA   2266566  165668.76   93039.48   258708.24
  2  BALTO/WASH INTL        BALTIMORE          MD    442042   18041.52   19722.93    37764.45
  3  LOGAN INTL             BOSTON             MA    954958  127815.09   29785.72   157600.81
  4  DOUGLAS MUNI           CHARLOTTE          NC    707695   36242.84   15399.46    51642.30
  5  MIDWAY                 CHICAGO            IL    354704    4494.78    4485.58     8980.36
  6  O'HARE INTL            CHICAGO            IL   2563638  300463.80  140359.38   440823.18
  7  DALLAS/FT WORTH INTL   DALLAS/FT WORTH    TX   2289926  142660.95   86706.76   229367.71
  8  LOVE FIELD             DALLAS/FT WORTH    TX    288283    2216.70     242.87     2459.57
  9  STAPLETON INTL         DENVER             CO   1196183   67345.75   38043.73   105389.48
 10  DETROIT CITY           DETROIT            MI     36265     258.08       0.00      258.08
```

```
Descriptive statistics for FAA data

Variable   N         Mean      Std Dev      Minimum      Maximum
-----------------------------------------------------------------
PASSNGER  134   316265.75    459666.53    3.0000000   2563638.00
TOT_TONS  135    45050.85     95745.77    7.9500000    627392.49
-----------------------------------------------------------------
```

```
Distribution of Airport States

                              Cumulative  Cumulative
ST   Frequency   Percent     Frequency     Percent
---------------------------------------------------
CA          14      10.4            14        10.4
FL          13       9.6            27        20.0
TX          13       9.6            40        29.6
NY           7       5.2            47        34.8
OH           6       4.4            53        39.3
MI           5       3.7            58        43.0
```

Recap

This chapter has presented a quick orientation to the SAS language. It has described the organization of SAS datasets; discussed the distinction between DATA steps and PROCs; presented general syntax rules; discussed assumptions that SAS makes for you; and given an overview of the different styles of program execution. The concepts were reinforced by the presentation of a small program. In the next chapter, we will discuss the way to prepare your data for analysis.

Preparing the Data for Analysis

3

SAS programming revolves around the creation and analysis of SAS datasets. You may be provided with one or more SAS datasets for your classroom or professional use. Very often, however, you need to construct a SAS dataset from raw data. SAS provides a wealth of tools for the reading of raw data into SAS datasets, a process often referred to as *importing*. This chapter presents some of the SAS language's more commonly used features for dataset creation. It also points out syntactical and logical mistakes frequently made by those starting out in SAS programming.

The discussion is more easily understood if we step back from the particulars of SAS syntax and discuss the task in general terms. To successfully read a raw dataset, you need to know several things and answer several questions. What will you call it? Where is it located? (That is, what is its file name?) Also, how are the data laid out? Is each variable in the same place in every record, or are the variables separated by one or more spaces? How many lines of input data in the raw dataset are there for each observation? What is the "unit of observation"? In many cases it's obvious: respondent to a survey instrument, geographic unit such as a county, and so on. Sometimes there are important subtleties: for example, the observation may be a combination of patient and visit rather than simply the patient.

We will soon see that there are SAS statements that address these questions. The DATA statement begins the DATA step. It names the SAS dataset you are creating. The INFILE and CARDS statements can be used to identify the location of the raw data. The INPUT statement describes the layout of the raw data. Finally, the LIBNAME statement tells SAS where to store the SAS dataset. The rest of this chapter will discuss their use in more detail.

Understanding the Input Data

As noted above, before writing your program you should consider some of the data's characteristics. A fundamental issue is the *layout* of the lines of data. If you are using an external data source, such as U.S. Census or survey data, you probably have access to a written description of the layout. These "codebooks" typically contain layout information along with the valid values of each variable, comments about their usage, and other helpful information. Careful study of the "codebook" before using the data always saves time and avoids headaches and confusion down the analytical road!

In most data files, each variable occupies the same column location(s) in each line or "record." You should know the beginning and ending columns of each variable you want to read. Note that SAS does not care if there are no empty columns between variables. A variable could use columns 12 through 14 and the next variable of interest could be in columns 15 through 16. Such compact, space-saving formats are common and, unfortunately, difficult to "eyeball," or casually inspect. Another layout issue is the number of data lines per unit of observation. Usually there is only one line per unit, but sometimes multiple lines are used. The INPUT statement has features that let you identify the line to read from in each unit of observation.

Another data issue is that of *data type*. As we noted in Chapter 1, SAS recognizes two types of data: numeric and character. Numeric values are, as their name suggests, representations of numbers. Items that will be arithmetically manipulated should be identified to SAS as numeric data types (this is accomplished in the INPUT statement, described later in this chapter). Common numeric variables are salaries, test scores, and ranks. Numeric data can, in some circumstances, be entered with seemingly nonnumeric values. For example, "3E2" is valid scientific notation and "5,100" is valid, provided that you supply the correct instructions in your program.

The other data type, character, is less restricted in content. It can contain *any* value from 1 to 200 characters in length. Names and addresses are examples of frequently used character data. Bear in mind that SAS stores character data exactly as it reads it from the data source. Blanks and case, which may be subconsciously ignored as you read the data, are significant to SAS. The following lines are different to SAS's "eye," even though they appear identical as you read them:

```
SAS Institute
SAS  Institute
SAS institute
```

Missing Data

Incomplete data crop up even in the most controlled situations. Such *missing data* are common, and are easily dealt with by SAS. The way you indicate missing data to SAS depends on the method of input you use. If you read specific columns for a variable and encounter only blanks or a period (.), SAS considers the variable to be missing for the observation. The value is stored by SAS as the special character "." If you are reading data separated by blanks (list input, described below), you must explicitly enter the "." Keep in mind that the number 0 is *not* a missing value — it is simply a measurement of 0 for a variable. The impact of missing values on calculations and statistical procedures can be significant and is discussed throughout the remainder of this book.

Both character and numeric data supplied from external sources may not follow SAS's rules for representing missing data. Values of 97, 98, and 99, for example, may indicate "refused to answer," "unclear," and "inappropriate" for a survey question. If these are all missing values for the purpose of your analysis, you need to convert each one to a SAS missing value. This is easily accomplished by using IF-THEN statements, described in the next chapter.

Naming the Dataset: The DATA Statement

Once you have an understanding of the unit of observation, the layout of the data, the data types, and the representation of missing values, you are ready to write your program. The first piece of information that SAS requires about the dataset you are building is its name. This is given using the DATA statement. DATA also serves as the signal to SAS that a DATA step is beginning. The syntax is straightforward:

```
DATA dataset_name;
```

`dataset_name` is the name you give to the dataset. It can be up to eight characters long, can contain alphabetic characters, numbers, or underscores (_), and cannot begin with a number. Ideally, it should give any reader of the program an idea of the dataset's content: names such as `fiscal93` and `counties` are better than `thing` and `mydata`, for example.

Locating the Data: The INFILE, CARDS, and DATALINES Statements

The next step in the dataset-building process is to identify the location of the raw data. The data can be stored in a host of formats and locations. Typically, however, you will use either *in-stream data* (stored in the same file as your program) or data stored in an external file. In-stream data are, as the name implies, lines of data contained within the program itself. This makes the programming effort a little easier to keep track of, since the data and the program are in the same place. It becomes impractical, however, when large amounts of data are used. The start of in-stream data is indicated by the CARDS or DATALINES statement. Their syntax is to the point:

```
CARDS;
DATALINES;
```

The CARDS or DATALINES statement should be placed at the end of the DATA step. The end of the in-stream data is signaled by any line containing a semicolon (;). The RUN statement is often used for this purpose, as illustrated here:

```
CARDS;
one or more lines of data go here
RUN;
```

The alternative to in-stream data is use of external files. These are separate, nonprogram files, which contain only data. Use the INFILE statement to identify the external file. The general form of the statement is

```
INFILE "file_location";
```

Its format will vary according to the operating system you are using. Your instructor or company help-desk staff can show you the appropriate format for your computer. You generally place it after the DATA statement and before the INPUT statement, discussed later in this chapter. Here are some common uses:

Exhibit 3.1

Sample INFILE Statements

Operating System	Example
MVS	`INFILE "sys27514.master.data"` `INFILE "sys27514.project.data(fy94)"`
VMS	`INFILE "[growth.rawdata]rural.1993to1995"`
UNIX	`INFILE "/home/frank/growth/rur9395.dat";`
DOS, Windows, OS/2	`INFILE "c:\growth\raw\rur9395.dat"`

When you are first trying to read the data, you may want to test the program on only part of the raw data. To do this, use the OBS option, as shown here:

```
INFILE "c:\growth\raw\rur9395.dat" OBS=100;
```

In this example, `OBS=100` tells SAS to make only the first 100 lines available for processing. If the raw dataset `rur9395.dat` contained fewer than 100 lines, SAS would simply process those that were available. No WARNINGs or NOTEs would be printed in the Log.

Describing the Data Layout: The INPUT Statement

At this point in the program, SAS knows what to name the dataset (DATA statement) and where the data are located (INFILE or CARDS statements). The remaining task is to describe the way to move through the dataset and read the variables. This is done using the INPUT statement. Because it can read virtually any form of data, the INPUT statement's syntax is probably the most complicated in the SAS language. Some of its most simple and powerful elements are reviewed in this section.

The INPUT statement can take several forms. All forms communicate the same information to SAS: the name of the variable to be read, how it will be stored (character or numeric data type), the columns to read the data from, and, optionally, the number of decimal places for a numeric variable.

List Input

The first form of the INPUT statement we discuss is *list input*. Data in this format are entered without regard to column location. Two restrictions are placed on the data: First, variable values must be separated by one or more blanks. Second, there must be as many values in each line of data as there are variable names in the INPUT statement. Such free-form data entry is often found when data are in-stream and contain relatively few observations. Its syntax is the simplest form of INPUT described in this chapter:

```
INPUT var <$> ...;
```

In this example,

- ❑ *`var`* is a variable name. It follows the same naming rules as names for SAS datasets, described in the preceding section on the DATA statement.
- ❑ `$` identifies *`var`* as a character variable. It never follows numeric variables.

Look at the following program:

```
DATA carolina;
INPUT code st $ rank;
CARDS;
100 NC 2
 100  NC  2
100      NC 2
    100 NC 2
RUN;
```

SAS list input is very forgiving of erratic and inconsistent spacing in the data. The extraneous blanks will be ignored and the variables will be read correctly. The values of the variables will be identical in all four observations.

Column Input

Most external files, particularly commercially available data, have a regular pattern for data location. Column input is one of two methods you can use to read such data. Unlike input of list data, it requires that you know the beginning and ending columns of each variable. The actual data value may be read anywhere within the column range. The syntax is basically an elaboration of list input:

```
INPUT var <$> start <-end> <.dec> ...;
```

In this example,

- ❑ `var` and $ have the same meaning as in list input.
- ❑ `start` is the column number in the dataset that is where the data value for the variable begins.
- ❑ `-end` is the column number in the line where the variable values end. If the variable is only one column wide, you don't have to specify `-end`.
- ❑ `.dec` tells SAS how many implied decimal places there are in the field. Decimal points actually entered in the data override the `.dec` specification.

Column input is demonstrated in the following program (the column ruler following the CARDS statement is not part of the data). Note that data values do not have to be right-justified within the specified columns.

```
DATA testscor;
INPUT init $ 1-3  group 5-6  score 8-12;
CARDS;
```

```
              1
1---5----0---
fcd  1  33.4
kah 2    35
joh  1  38
```

Formatted Input

Column-style input statements are suitable for reading most files. In some cases, however, the data may more easily be read with *formatted* input. Like the column style, formatted input requires that you know the location of the beginning of the variable in the line of data as well as how many columns it uses. It also uses a powerful SAS feature called *input formats*, or *informats* for short. These are instructions that tell the SAS System how to read raw data. SAS comes with a wide variety of formats that help you read virtually any kind of data. We'll touch on a few of them in this section. But first, let's take a look at the syntax of the INPUT statement using formatted input:

```
INPUT </> <@col> var fmt. ...;
```

In this example,

- / tells SAS to go to the next line of the raw data before reading the next variable. This gives you the ability to read a single observation's data from more than one input record. Be careful when you use this feature! If the raw data have, say, three lines of data per observation, you should have two /'s in the INPUT statement. If you do not, SAS will read the data incorrectly and create a dataset full of gibberish.
- `col` is the column number at which SAS begins to read. If the `@col` specification is omitted, SAS reads from the next column in the input data.
- `var` is the name of a variable.
- *fmt.* is a format specification. These are discussed below. Note the trailing period. This is required to avoid confusing the format with a variable name.

INPUT Statement Formats

Some of the more popular input formats are shown in Exhibit 3.2. You should be aware that the SAS System comes with *many* others and that you can, with a little effort, write your own. Such user-written formats are discussed in Chapter 13.

Exhibit 3.2

Numeric INPUT Statement Formats

Format	Description	Width Min/ Max	Raw Data	INPUT Specs	Numeric Value Stored
`w.`	Numbers, without decimals.	1/32	230	3.	230
`w.d`	Numbers, with decimals (*d* cannot exceed *w*).	1/32	1000 10300 10.12	4.2 5.3 5.0	10 10.3 10.12
`COMMAw.d`	Embedded commas, dollar signs. Use 0 or 2 decimal places.	1/32	4,522	comma6.	4522

Formatted input is demonstrated in the following example (the column ruler following the CARDS statement is not part of the data):

```
DATA drug1;
INPUT id $3. @5 gender $1. @8 salary comma8.;
CARDS;
         1         2
----+----0----+----0
0A1 M  45,000
0A2 M    34,800
0B1 F  49,000
```

Mixing Input Styles

The different input styles give you considerable flexibility when reading your data. If you still feel constrained by the syntax, remember that you can use more than one style in a single INPUT statement. For example, column input could be used for some fields, then a COMMA format used for data entered with embedded commas. Mixed input styles are demonstrated in the following example (the column ruler following the CARDS statement is not part of the data):

```
DATA drug1;
INPUT id @5 race $1. score1 score2 dept 18-20;
CARDS;
         1         2
----+----0----+----0
1   W  88 90      100
 2  B     87 100  101
12  W  90    90   101
```

This example is not terribly easy to follow. It is usually easier to read a statement that uses only one input style. The example is presented primarily to show that combining INPUT styles is syntactically valid.

Which Style to Use?

The INPUT style you select is based partly on the characteristics of the data and partly on preferences you develop over time. List style is usually acceptable if you need to enter small amounts of data (few observations and/or few variables). Column and formatted styles are usually required if the data are being provided to you by external sources, such as government agencies. These styles also allow you to skip over variables present in the raw data that you don't want to read. Such bypassing is not feasible with list input.

Common Problems

Problems in reading raw data generally fall into two categories: bad data and incorrect specifications for reading the data. When SAS encounters problems with the data, it *sometimes* issues messages in the Log, noting which variable in which line of the dataset caused the problem.

Data problems include

- ❑ *Typos.* The letter "O" is used instead of the number "0". The letter "l" is used instead of the number "1".
- ❑ *Coding errors.* Male/female codes are entered as M/F rather than the expected 1/2.
- ❑ *Mixed case.* Case matters in data! If values for a variable are entered as "WC" in one observation and "wc" in another, SAS considers them to have different values. This may or may not be acceptable from an analytical standpoint. It is *not* considered a problem by SAS.

Specification problems include

- ❑ *Incorrect location.* INPUT attempts to read a variable from columns 20 through 25 when, in fact, it should have read from columns 19 through 25. This may or may not create an error! If the Log shows an error for a variable for every observation read, you've probably misspecified the column locations.
- ❑ *Confusing columns and formats.* Another common mistake is leaving off the period (.) at the end of a format. In the example below, the first INPUT statement reads the first four columns of a record; the second, without the period, reads *only* from column 4! They are both valid from a syntax point of view, but can produce greatly different results. You may be alerted to this problem by seeing unexpected numbers in listings or analyses.

```
INPUT income 4.;
INPUT income 4;
```

- *Bad data type.* You expected a numeric value for, say, a yes/no question, but instead were supplied with character data. That is, instead of 1/0 coding, the data contained Y/N.

- *Incorrect format.* The data contained commas in a field, but you didn't use the COMMA format.

Let's assume that you've read the data correctly. Rather than have SAS go through the slow process of reading the raw data every time you run the program, it is sometimes easier to create a permanent SAS dataset. This task is discussed in the next section.

Permanent Versus Temporary SAS Datasets

As you might infer from the preceding sections, the process of writing the INPUT statements to read raw data can be tedious, especially if more than one program is going to be reading the data. An effective way to avoid respecifying the data is to create a *permanent SAS dataset.* The datasets you've seen so far have been temporary. They were deleted once your batch job was completed or your interactive session ended. Permanent datasets remain in your computer after the job or session ends. You may use them later without having to reread the raw data.

Using permanent datasets is easy to do and saves computer resources. This section discusses what's in SAS datasets, describes a new feature of the DATA statement, and introduces the LIBNAME statement, your program's link to the rest of the computer system.

What's in the Dataset?

In Chapter 2, we noted that SAS programming revolves around the creation and manipulation of SAS datasets, and that PROCs will operate only on these datasets. What exactly are SAS datasets, besides being the only ones that work with SAS? They have a few special features. First, they contain *both* your data values *and* information for SAS about how to read them. Thus they are self-contained. This means you don't have to use a DATA step to read your data once it is stored as a SAS dataset. The original data plus any calculations you may have performed are stored in the dataset.

Second, SAS datasets are very efficient. They often require less storage space than the raw data from which they came. They are also written so that they move data into PROCs very quickly. The SAS dataset is, in effect, a ready-to-use collection of data and information about the data. This is in contrast to raw data, which must be processed by a DATA step before it can be used by PROCs.

In addition to the data values and the instructions for reading them, the SAS dataset contains other useful, optional information. *Labels*, extended descriptions of the variable names, are stored in the dataset. *Formats* and informats used for reading and displaying the data are stored for each variable as well. The *sort order* of the dataset, the order in which observations are arranged, is also stored.

Linking to the System: The LIBNAME Statement

The datasets used so far in this book have had an important limitation. They were deleted once the program was completed. If we want to store the dataset permanently and have it available for future programs, we must use a LIBNAME statement to identify the location where the dataset will be stored. Its syntax is shown below (keep in mind that some extensions to this syntax may be required on your computer system):

```
LIBNAME libref "location";
```

In the above statement,

- `libref` is a name you assign to the location, a sort of "handle" or nickname. It must start with a letter, and may contain up to eight letters, numbers, and underscores (_). It must be placed before the first DATA statement that refers to it.
- `location` identifies, in IBM MVS systems, the name of the system file holding the SAS datasets. In other systems, such as DOS, OS/2, UNIX, and VMS, `location` names a directory that will contain one or more SAS datasets.

It's important to understand that the location to which LIBNAME points does not actually name a SAS dataset. All it says is, in effect, "Here's an area on the computer system's disk. You may find *any number* of SAS datasets in this location." These locations are sometimes called "libraries." To illustrate this, consider a DOS system and the command

```
dir c:\emission\sasdata
```

Part of the system's reply looks like this:

```
se93.ssd
se94.ssd
readme.txt
ne93.ssd
```

SAS uses the `ssd` extension (the part following the period) in DOS file names whenever it writes a SAS dataset. This varies according to the operating system you're using. The first part of the file name is the name of the SAS dataset. Thus three of the four files in the directory are SAS datasets, and their names are `se93`, `se94`, and `ne93`. A single LIBNAME, such as

```
LIBNAME toxic 'c:\emission\sasdata';
```

could refer to three SAS datasets. The remaining piece of the puzzle is how to communicate to SAS that you want to use the `toxic` libref. This is easily done by adding an extension to the SAS dataset name. This is discussed in the following section.

"Two-Level" Dataset Names

If you scan the sample output in Chapter 2, you'll see NOTEs in the Log such as "The data set WORK.SAMPLE has 24 observations...." WORK is a special libref automatically assigned by SAS when your program begins to run. Even though you enter TEMP as a dataset name in the DATA statement, SAS adds WORK.

You use the libref assigned in the LIBNAME by referring to a two-level dataset name of the form

```
libref.name
```

where *`libref`* was assigned in an earlier LIBNAME statement and *`name`* is the name of the SAS dataset you are reading or writing. To complete the example started earlier,

```
LIBNAME toxic 'c:\emission\sasdata';
DATA toxic.se95;
INFILE 'c:\emission\rawdata\stheast.1995';
INPUT ...;
```

The two-level name `toxic.se95` tells SAS to write the dataset `se95` to the directory pointed to by the libref `toxic`. The LIBNAME statement containing `toxic` tells us that this location is the subdirectory `\emission\sasdata`.

Recap

Syntax Summary

This chapter has covered the following DATA step syntax. (See the syntax description in Chapter 1 for an explanation of the notation.)

```
LIBNAME libref "location";
DATA dataset_name;
<INFILE "file_location"; | CARDS;>
INPUT var <$> ... ;
INPUT var <$> start <-end> <.dec_places> ...;
INPUT </> <@col> var fmt. ...;
<CARDS; | DATALINES;>
```

Remember that multiple INPUT styles (list, column, and formatted) may be combined in a single INPUT statement.

Content Review

Before you can analyze your data, you have to be able to read your data. This chapter has outlined the relatively few statements required to translate raw data into a SAS dataset. We presented the syntax and usage of the DATA, INFILE, and INPUT statements. The INPUT statement warrants special attention. It enables you to read a wide variety of data and thus has different formats and options to suit your needs. We discussed the different INPUT styles and suggested how to choose among them. SAS datasets can be stored and reused from program to program. This is accomplished by using the LIBNAME statement and two-level SAS dataset names.

Now that you are able to create and save SAS datasets, you can either use them in procedures or further manipulate them in DATA steps. If you want to begin to use them in procedures, skip to Chapter 6. If you want to learn to perform calculations, take subsets of the data, and combine two or more SAS datasets, continue to the next chapter.

Introduction to DATA Step Programming

4

In earlier chapters, we discussed how to read raw data, create a SAS dataset, and use the dataset in some simple and useful procedures. These activities are common to virtually all research and analysis applications. Now you are faced with a simple decision: Do you start by analyzing the data or do some programming before it's ready to use?

This chapter addresses the "programming" part of the question. It describes the use of some simple DATA step programming tools. Among these are

- ❑ Comments, for documenting your work
- ❑ The SET statement, for reading permanent SAS datasets
- ❑ Assignment statements, for creating and modifying variable values
- ❑ IF-THEN statements, for performing actions based on logical conditions
- ❑ The OUTPUT statement, for writing an observation to the dataset
- ❑ The FORMAT and LABEL statements, for enhancing the display of variable names and values

If your dataset is ready to analyze, skip to Chapter 7 for a description of statistical PROCs. Keep in mind, though, that eventually you will need at least some of the DATA step features discussed here.

Documenting Your Work: Using Comments

One of the most basic, and most frequently neglected, aspects of programming is documentation. It does not have to be elaborate. Indeed, its presence in your program is entirely optional. We mention it here because it is both good programming and research practice. If you use SAS or any other programming language professionally, documentation becomes essential. The sooner you get into the habit, the better.

The content of the documentation is up to you. It should include anything that will help you remember what tasks the program is supposed to perform. Items you may want to include are

- Name(s) of dataset(s) used by the program
- Which, if any, program(s) must be run before or after this program
- Special notes about why a calculation is performed a certain way
- Overall objective of the program

A good rule of thumb is that if you find yourself writing notes to yourself about the program on scraps of paper, those notes should probably be entered in the program itself.

Use comments to document your programs. You have two forms from which to choose. The syntax is straightforward in both cases:

```
/* comment text goes here  */
* comment text goes here ;
```

The "/*" indicates the start of a comment; the "*/" ends the comment. The only caveat is that users in IBM MVS environments should not begin a "/*" comment in column 1. Comments may extend across more than one line:

```
/* REG1.SAS
   This program uses the GEO dataset and
   performs logistic regressions. Models as
   per meeting with committee April 2, 1995.
*/
```

"/*" comments may be entirely contained within a statement:

```
xprod = (yr1 * rt1)   /* 1st yr not adjusted */
      + (yr2 * rt2 * 1.01); /* Yr 2 adjuested
                               up by 1% */
```

Be careful not to overcomment — don't point out the obvious:

```
* We do some real tricky stuff in this section ;
DATA version1;    /* Begin DATA step for VERSION1
*/
```

Such comments do not contribute to an understanding of the program's logic and purpose.

Reading SAS Datasets: SET

SAS datasets were discussed in earlier chapters. We saw how to create them and save them in permanent dataset "libraries" for subsequent programs. If you want to use a SAS dataset in a DATA step, you must be able to access the data. This is done with the SET statement, whose function closely resembles that of the INPUT statement. The difference is that it operates on SAS datasets. INPUT reads a line of data from a raw data file, converts it to SAS data format, and makes it available for processing by DATA steps or PROCs. SET reads an observation from a SAS dataset and makes it available for processing by DATA steps or PROCs.

Syntax

The format of the SET statement is

```
SET dataset_name;
```

Here, `dataset_name` is the one- or two-level name of the dataset. You can specify options that control the amount of data to read:

```
SET dataset_name(OBS=n KEEP=keep_vars);
SET dataset_name(OBS=n DROP=drop_vars);
```

In these statements,

- ❑ `OBS=n` instructs SAS to use only the first *n* observations from `dataset_name`. This option is useful when you are developing a program. Run the program with a small value of `n`. Then, when the code appears correct, remove the option and process all observations.
- ❑ `KEEP=keep_vars` tells SAS to read only select variables from `dataset_name`.
- ❑ `DROP=drop_vars` tells SAS to read all variables except `drop_vars` from `dataset_name`. Remember, we use the SET statement to *read* a dataset — the variables in the `drop_vars` list are still stored in the data-

set. By using the DROP option, we are just making them unavailable for use in the DATA step.

If a variable is specified in both a DROP and a KEEP list, the variable is dropped. Judicious use of DROP or KEEP options can greatly speed up the processing of SAS datasets with large numbers of variables, especially when you only need to read a few variables from the dataset.

Examples

The following examples demonstrate the use of the SET statement and show how it can affect the dataset being written. Keep in mind that in complete programs, other DATA step statements would follow these program excerpts.

Task: Create a dataset from the first 20 observations of a permanent SAS dataset.

```
LIBNAME sasdata
'user:[fcd7484.trans.sasdatasets]';

DATA first20;
SET sasdata.fy9394(OBS=20);
```

Task: Overwrite a permanent dataset. (Possibly a mistake, since only 20 observations from `fy9394` are being used!)

```
LIBNAME sasdata
'user:[fcd7484.trans.sasdatasets]';

DATA sasdata.fy9394;
SET sasdata.fy9394(OBS=20);
```

Task: Create a temporary dataset containing selected variables and observations from a permanent dataset.

```
LIBNAME sasdata
'user:[fcd7484.trans.sasdatasets]';

DATA sasdata.subset94;
SET sasdata.fy9394
    (OBS=100 KEEP=id loc bal1 bal2);
```

Task: Create a permanent dataset based a subset of the input dataset's observations.

```
LIBNAME sasdata 'user:[fcd7484.trans.sasdata]';

DATA sasdata.hibal94;
SET sasdata.fy9394(KEEP=id loc bal1 bal2);
IF bal1 > 100000 or bal2 > 100000 THEN OUTPUT;
```

Performing Calculations: Assignment Statements

One of the most common data manipulations is the creation or modification of variables. You may, for example, want to create indicators, or *flags*, showing whether an observation falls into a high-, medium-, or low-income group. You may want to transform an existing variable from, say, pounds to kilograms. The range of applications is boundless, and they can readily be coded by using assignment statements.

Syntax and Usage Notes

The format of the assignment statement is straightforward:

```
var = operand operator operand ... ;
```

In the above,

- `var` is the variable being created. It is sometimes referred to as the *target*. It may be the name of a variable already in the dataset or a new name.
- `operand` is either a constant or a variable.
- `operator` is an arithmetic operator. These are + (addition), - (subtraction), * (multiplication), / (division), and ** (exponentiation).

The syntax is simple but powerful. Any number of operands and operators may be strung together to form complex calculations. Parentheses may be used to clarify the relationships between the different pieces of the calculation. Operations within parentheses are performed before operations outside the parentheses. This is demonstrated in the examples below. If parentheses are absent, exponentiation is performed first, followed by multiplication, division, addition, and subtraction. Similar operations are performed left to right. Rather than have to remember these rules, it's usually easier to include parentheses to make the intent of the calculation clear. Keep in mind the following caveats about calculations:

- *Order is important.* SAS is always aware of errors in syntax, but is oblivious to problems in logic. If you *use* a variable in an assignment statement before you *define* it, the target variable will be missing. Consider the following:

```
DATA roster;
height = (12 * feet) + inches;
INPUT name $ 1 - 15 feet inches;
```

The syntax of each statement is correct. The logic, however, is flawed. We use the variables `feet` and `inches` in the calculation *before* we read them in the `INPUT` statement. As a result, when the assignment statement is executed, `feet` and `inches` are missing. This leads to our next point.

- ❑ *Missing values are important.* If any operand is missing, the result, or target, is set to missing. SAS will display a note in the Log indicating where this happened and how many times.

- ❑ *Overwrite values carefully.* The same variable may be used on both sides of the equal sign. Keep in mind that the original value is overwritten and may not be possible to get back (unless, of course, you recreate the original dataset). Look at the following statements:

```
sal = sal * 1.05;
pay = pay * adjust;
```

Original values of `sal` can be computed because it is multiplied by a constant. We are not so fortunate with `pay`. Since values of `adjust` can vary from observation to observation, there is no way to determine the original values of `pay`. Here, and probably with `sal` as well, it is probably better to create new variables rather than to replace variables, as follows:

```
newsal = sal * 1.05;
newpay = pay * adjust;
```

- ❑ *Consider using functions.* The SAS System has many built-in tools for performing calculations including univariate statistics, financial calculations, logartihmic transformations, and so on. These tools are also more robust than assignment statements when handling missing values. These "functions" will be discussed in Chapter 11.

Examples

The examples in this section demonstrate not only correct syntax but good programming style. Note how liberal use of blank space, parentheses, and multiple lines clarifies the calculation. In some cases, an assignment is deliberately followed by its poorly written equivalent to emphasize the benefits gained from good style.

```
newscore = (score1 * 1.05) + (score2 * .98);
loan_val = prin * (1+r)**time;
loan_val = prin * 1 + r**time;
residual = observed - expected;
```

```
/* Good: */
xprod = (r1 * d1 * sc2) +
        (r2 * d2 * sc2) +
        (r3 * d3 * sc3) ;
/* Poor: */
xprod = (r1*d1*sc1) + (r2*d1*sc2)
  + (r3*d3*sc3);
```

Notice the difference the parentheses make in the two versions of `loan_val`. In the first, the sum of `1` and `r` is raised to the power of `time`. In the second statement, without the parentheses, only `r` is raised to the power of `time`.

Building Logical Expressions

As we saw above, the syntax of the assignment statement is simple, yet powerful enough to satisfy most computational needs. There are situations, however, where a calculation needs to be handled differently depending on the values of one or more variables. For example, pay raises may need to differ according to department. Many types of activities, not just calculations, may need to be performed depending on the values and relationships of one or more variables and constants.

Before examining the DATA step tools for doing this conditional execution, we need to discuss logical expressions. These are the SAS equivalent of English phrases such as "the value of the state variable is either New York or Oklahoma" and "salary exceeds $50,000 and pay grade is less than 50." These expressions are the building blocks of the IF-THEN-ELSE statements described in the next section. They are also used in the WHERE statement, discussed in Chapter 6.

Comparison Operators

The basic part of any expression is a comparison between two variables and/or constants. The comparison is performed for each observation. The result is either true or false. Exhibit 4.1 describes comparison operators, the tools used to specify the type of comparison.

Exhibit 4.1

Comparison Operators

Operator[1]	Meaning	"True" when
`EQ =`	"equals"	The value on the left of the operator is equal to the value on the right.
`GT >`	"greater than"	The value on the left is larger than that on the right. Character data are evaluated alphabetically. Be aware, though, that nonalphabetic character data are handled differently in different operating systems.
`LT <`	"less than"	The value on the left is less than that on the right.
`GE >=`	"greater than or equal to"	The value on the left is at least as large as that on the right.
`LE <=`	"less than or equal to"	The value on the left is, at most, equal to the value on the right.
`IN`	"in"	The value on the left is in the list of values specified on the right of the operator. The values are entered within parentheses, such as `(1, 2, 5, 6)` or `("AU", "NZ")`.
`NOTIN`	"not in"	The value on the left is not found in the list of values specified on the right of the operator. The list syntax is the same as that of IN.

[1]The character and symbolic forms are equivalent. For example, EQ and = may be used interchangeably.

Note that numeric missing values will compare low. If variable x is missing in an observation and we use the expression

```
x < 42
```

the comparison will be true, since a missing value is less than 42. If the comparison were coded

```
x > 42
```

the result would be false, since a missing value is not greater than 42.

Logical Operators

Very often the simple syntax described above is not adequate. You may want to perform an action depending on the truth of more than one comparison. These situations require the use of logical operators forming *compound expressions*. The usage of the two most commonly used operators is explained in Exhibit 4.2.

Exhibit 4.2

Using Logical Operators

<table>
<tr><th>Operator</th><th>Explanation</th></tr>
<tr><td><code>comp1 & comp2</code>
<code>comp1 AND comp2</code></td><td>Both <code>comp1</code> and <code>comp2</code> must be true for the action to take place.</td></tr>
<tr><td><code>comp1 | comp2</code>
<code>comp1 or comp2</code></td><td>Either <code>comp1</code> or <code>comp2</code> must be true for the action to take place.</td></tr>
</table>

Notice your choice of notation: the ampersand (`&`) may be written as `and`, and the vertical bar (|) may be written as `or`. Any number of conditions may be linked with the two operators. Parentheses must be used to clarify which part of the comparison should be done first (this is shown in the following examples).

Examples

Exhibit 4.3 presents simple and compound expressions along with a description of each expression's effect.

Exhibit 4.3

Interpretation of Expressions

<table>
<tr><th>Expression</th><th>True when ...</th></tr>
<tr><td><code>st = 'ME'</code></td><td>state equals '<code>ME</code>'. Notice that a <code>st</code> value of '<code>me</code>' would not be used — character comparisons are case-sensitive!</td></tr>
<tr><td><code>grosspay > 50000</code></td><td>gross pay exceeds 50000. Missing values, and those equal to 50000, would compare false.</td></tr>
<tr><td><code>grosspay >= 50000</code></td><td>gross pay is 50000 or more.</td></tr>
<tr><td><code>st in ('ME','me')</code></td><td>state is either uppercase or lowercase 'me'.</td></tr>
<tr><td><code>id = . | name = ' '</code></td><td>either <code>id</code> or <code>name</code> is missing.</td></tr>
<tr><td><code>id = . & name = ' '</code></td><td>both <code>id</code> and <code>name</code> are missing.</td></tr>
<tr><td><code>(grd <= 20 & sal > 40000)</code>
<code>| (grd >= 40 & sal < 60000)</code></td><td>either of two conditions holds: <code>grd</code> is less than or equal to 20 and <code>sal</code> is over 40000 or, second, <code>grd</code> is at least 40 and <code>sal</code> is less than 60000.</td></tr>
</table>

Conditional Execution: IF-THEN-ELSE

The DATA steps shown so far have simply executed the statements in a given step sequentially. That is, each statement in the DATA step was executed for every observation. There are situations where this is not desirable. You may want to create new variables based on values of one or more other variables, for example. Saying this in prose rather than in code highlights the underlying process: "If age is less than 21, premium equals base times 1.10. If age is between 21 and 25, premium equals base times 1.08. Otherwise, premium equals base times 1.04." This logic is implemented in SAS using IF-THEN-ELSE statements.

Forms of IF-THEN-ELSE

Here are several forms of IF-THEN-ELSE syntax:

Form 1:

```
IF comparison1 THEN action1;
```

Form 2:

```
IF       comparison1 THEN action1;
   ELSE                   action2;
```

Form 3:

```
IF              comparison1 THEN action1;
   ELSE IF comparison2 THEN action2;
   ELSE                     action3;
```

In the above, `comparison` is any logical condition. `action` is any executable SAS statement (assignment statements are often used here).

The ELSE statement instructs SAS to test another `comparison` if the preceding `comparison` was not met. An ELSE statement at the end of the statements can omit the `comparison`. This means that the associated `action` will be executed. Such an "unconditional" ELSE is a good way to create a catchall, or residual-class, variable. Only one "unconditional" ELSE is allowed per IF-THEN sequence.

Examples

Pay particular attention to the use of blank space. Indentation of ELSE statements makes the logic of the entire IF-THEN-ELSE sequence easier to follow. Here is the IF-THEN-ELSE implementation of the logic described at the beginning of this section:

```
IF            . <  age <  21 THEN due = base * 1.10;
   ELSE if 21 <= age <= 25 THEN due = base * 1.08;
   ELSE                          due = base * 1.04;
```

Notice the coding of the condition in the first ELSE statement. This is a shorthand notation for saying "age between 21 and 25." Also notice the comparison ". < `age`" in the first IF statement. If this comparison had been omitted, observations with missing `age` values would be computed here, since a missing value is less than *any* nonmissing value.

```
IF           . <  age <  20 THEN due = base * 1.10;
   ELSE if 21 <= age <= 25 THEN due = base * 1.08;
   ELSE                         due = base * 1.04;
```

This demonstrates a common and often bedeviling problem. What if `age` is 20? It is not caught by the IF statement, nor by the first ELSE. It lets the statements "fall through the cracks," letting the second ELSE assign the value to PREMIUM.

```
IF sc1 NE . & sc2 NE . THEN tot = run1 + run2;
   ELSE                     tot = -1;
```

`tot` is certain to have nonmissing values. If both `run1` and `run2` are nonmissing, we compute `tot` as their sum. Otherwise, we set `tot` to -1. Without the ELSE statement, `tot` would be unassigned, or missing (.).

A Note About Recoding

One of the popular applications for IF-THEN statements is to recode variables. This can take several forms:

- *One-to-one mapping.* Data may be stored as a short, easily entered value, but displayed as a longer, more readily grasped character string. You may store a variable RACE with values "W" and "N", but want to display the categories as "White" and "Nonwhite".
- *Grouping.* Data often need to be collapsed into groups, forming what is sometimes referred to as a *many-to-one relationship*. The grouping may be driven by a specific research question or by natural, preexisting hierarchies. Interval scale measurements, such as test scores, may be collapsed into ordinal variables such as "high", "medium", and "low". They may also be grouped into numeric variables (3 rather than "high", for example). Likewise, states may be grouped into regions, or branch offices into districts.

IF-THEN-ELSE structures are appropriate for such recoding, but consider two important points. First, be aware of lengths when recoding to character values. The following program fragment illustrates this:

```
IF percent > 60 THEN density = "high";
   ELSE              density = "medium";
```

Observations falling into the ELSE category will have `density` values of `medi` rather than the expected value `medium`. The length of a character variable's first reference is the one SAS uses to store the variable. Thus the first encounter with `density` has length 4. All other values, even if they are longer than four characters, will be stored in four characters. The excess characters will be cut off.

There are several ways to circumvent this. One is the LENGTH statement, which is discussed in Chapter 14. Another is to write the values with trailing blanks as needed:

```
IF percent > 60 THEN density = "high  ";
   ELSE              density = "medium";
```

The second point about recoding is the availability of alternative methods. IF-THEN-ELSE coding is adequate for applications where the number of assigned categories is small. Once the number of categories increases, so does the possibility of introducing errors. A safe and relatively easy-to-use alternative is PROC FORMAT. This gives you the means to write your own formats, which you can then use to perform recoding. User-written formats are discussed in Chapter 13.

Selecting Observations: OUTPUT

Normally, a DATA step executes statements beginning with the DATA statement and continues until it reaches a RUN statement. At that time, it writes an observation to the dataset identified in the DATA statement. In some situations, you may want to take control over which observations are written. You may, for example, want only `RACE` values '`H`', '`B`', and '`O`' when `INCOME` is greater than 20000. Filter out unwanted observations from the dataset you are creating by using the OUTPUT statement:

```
OUTPUT;
```

To implement this example, use the following statement:

```
IF race IN ('H','B','O') & income > 20000
   THEN OUTPUT;
```

The OUTPUT statement can be used in more than one location in the DATA step. This means it can be the *`action`* of any number of IF statements' conditions.

When using OUTPUT, remember that the observation is *immediately* written to the dataset. Look at the following code fragment (the DO and END statements delimit the beginning and end of a logically related group of statements and are discussed in Chapter 11):

```
IF salary >= 50000 THEN DO;
   OUTPUT;
   sal_grp = "High";
   END;
```

The observation is written to the dataset (OUTPUT statement), then `sal_grp` is assigned. In effect, we have "zagged" before we "zigged," since the observation will always be written with missing values of `sal_grp`. This type of logic error is surprisingly common, is not flagged by SAS as an error, and is easily corrected by simply reversing the order of the assignment and OUTPUT statements.

Increasing Variable Description: LABEL

Normally, when SAS displays results, it simply prints the names of the analysis variables. This is somewhat limiting, since variable names are limited to eight characters. A handy, and optional, way to specify longer, more descriptive explanations of variable content is to specify variable *labels*. The format of the LABEL statement is:

```
LABEL variable = "description" ... ;
```

Here, *`description`* is up to 40 characters of text that explain the meaning of the variable. More than one variable can be specified in a single LABEL statement. Here is an example:

```
LABEL wgt      = "Gross weight (kilograms)"
      retlpric = "Retail price, (1988 US$)"
      ;
```

When a label is attached to a variable in a dataset, SAS will usually display it in PROCs. Although the use of labels is optional, you'll soon notice that when you work with a dataset long enough, you'll want the explanatory power they add to your output.

Controlling Display of the Data: FORMAT

By default, SAS chooses the most appropriate way to display character and numeric data. The choice is often inappropriate. You may want to see only two decimal places, display the first 10 characters of a long character variable, or insert commas and dollar signs to improve readability. The FORMAT statement addresses these (and other!) requirements. It associates variables and output, or display, formats in much the same way as the INPUT statement associates variables and input formats.

Before describing some of the SAS formats, let's take a look at some general features of their usage. First, display formats are just that — display only. Displaying a numeric variable with commas does not alter the way the data values are stored in the SAS dataset. Second, a variable can be displayed with, say, a COMMA format in one procedure and another — for instance, DOLLAR — format in a procedure later in the same program.

Third, formats require that you specify at least one and possibly two items: field width and decimal places. Field width is the number of columns or print positions used to display the variable. The minimum and maximum numbers of columns vary with the format. Decimal places are appropriate only for numeric variables. The number of places should not exceed the field width. If the field width is insufficient to display a value, SAS tries to use other formats and displays the following message in the Log:

```
NOTE: At least one W.D format was too small for
the number to be printed. The decimal may be
shifted by the "BEST" format.
```

The message is frustrating: it does not say *which* variable is in question or *how many* times the problem arose!

Exhibit 4.4 shows some of the more common display formats. Notice how the display is adjusted if you do not specify an adequate width ("w" value). Formatting features such as parentheses and commas will be removed first, so that the value may be displayed. SAS places priority on displaying the number.

Keep in mind that there are *many* other formats. Also remember that if a SAS format does not meet your needs exactly, you can create your own using the FORMAT procedure, described in Chapter 13.

Exhibit 4.4

Common Display Formats

Format	Min/ Max	Description	Stored	Usage Examples[1]	Displayed
COMMAw.d	2/32	Writes numeric variables with commas. *d* may be either 0 or 2.	1100 900 1100.1	comma5. comma5.2 comma8.2	1,100 900.0 1,100.10
DOLLARw.d	2/32	Like COMMA, but also inserts dollar signs ($).	1000 1000	dollar8.2 dollar9.2	$1000.00 $1,000.00
Ew.	7/32	Writes numeric variables in scientific notation	2888 -2 2.3	e7. e7. e8.	2.9E+03 □-2E+00 □2.3E+00
NEGPARENw.d	1/32	Displays negative numbers in parentheses. Commas are inserted where appropriate.	-1000 -1000 1200	negparen8.2 negparen11.2 negparen11.2	-1000.00 (1,000.00 □1,200.00
PERCENTw.d	4/32	Writes numeric values as percentages. Negative values are enclosed in parentheses.	1 1 .4 -2	percent6. percent6.2 percent6. percent6.	□□100% 100.0% □□□40% (200%)

Exhibit 4.4

Common Display Formats (continued)

Format	Min/ Max	Description	Stored	Usage Examples[1]	Displayed
SSNw.	11	Displays social security numbers with hyphens.	123456789 123 0 .	ssn11. ssn11. ssn11. ssn11.	123-45-6789 000-00-0123 000-00-0000 ...-..-....
$w.	1/200	Displays *w* leftmost characters.	Inactive □□test □□test	$10. $6. $4.	Inactive□□ □□test □□te
w.d	1/32	Displays numeric values with, optionally, *d* decimal places.	40.3 40.3 40.3 40.3 .1	5.0 5.1 5.2 5.3 3.1	□□□40 □40.3 40.30 40.30 0.1
Zw.d	1/32	Like w.d, but inserts leading zeros.	40.3 40.3	z5.0 z5.1	00040 040.3

[1] In the examples, blank columns are indicated by □.
[2] Minimum and maximum columns/print positions that can be used by the format.

Recap

Syntax Summary

This chapter has covered the following DATA step syntax. (See the syntax description in Chapter 1 for an explanation of the notation.)

```
/* comment text */
* comment text ;
SET dataset_name(<OBS=n> <KEEP=keep_vars>
                 <DROP=drop_vars>);
var = <(>operand operator operand <)> ... ;
IF condition1 THEN action1;
   <ELSE <IF condition2 THEN> action2;>
OUTPUT;
LABEL var = "description" ... ;
FORMAT var1 <var2 ...> format. ...;
```

Operators are an integral part of assignment, IF-THEN-ELSE and other statements discussed in this book:

Arithmetic:	`**, *, /, +, -`						
Comparison:	`=`	`>`	`<`	`>=`	`<=`	`IN`	`NOTIN`
	`eq`	`gt`	`lt`	`ge`	`le`		
Logical:	`&`	`|`					
	`and`	`or`					

Content Review

Once data have been read into a SAS dataset, there are many ways you can manipulate them in the DATA step. This chapter has reviewed some of the more commonly used DATA step statements. We discussed the use of SET to read an observation from a SAS dataset. The syntax and usage caveats of assignment statements were also presented. Some situations require that calculations be performed differently depending on values of one or more variables. IF-THEN-ELSE statements handle these tasks.

Sometimes an analytical task requires that you use a subset of the data. There are many ways to accomplish this. One of them, the OUTPUT statement, was described in the chapter. Finally, we presented two optional features used to increase the readability of the program: LABEL and FORMAT statements enhance variable description and data readability.

One of the most common data management tasks in the research process is combining datasets. The next chapter describes the most common techniques. Although the syntax is straightforward, the results can sometimes be different from what you intended. Chapter 5 points out some of the more common problems and how to avoid them. If all your research data are contained in a single SAS dataset with no need for periodic updates, skip Chapter 5 and read Chapter 6 for a discussion of how to use PROCs.

Combining Datasets

5

All of the DATA step programming discussed so far has been based on manipulation of a single SAS dataset. In many situations, using only one dataset is adequate: you may want to perform calculations, recode variables, extract subsets of the data, and so on. Sometimes, however, there is a need to combine data from two or more SAS datasets. Some common reasons for this follow.

- New surveys arrive in the mail. They need to be added to the dataset containing previously entered observations. The new observations can simply be appended to the dataset.
- New observations need to be added next to the old records. Continuing with the survey example, the new observations for each respondent need to be added so that all observations for a given respondent are next to each other. Wave 1 data for a respondent should be followed by data for Waves 2 and 3.
- Different sources need to be combined. Patient visit data need to be analyzed according to age groupings. Age is found on the demographic, or baseline, dataset. The visit dataset needs to be merged with the demographic dataset.

These and many other data-handling requirements are readily met with relatively few SAS statements. This chapter discusses a few of the more common dataset-combining techniques. We will examine new uses of the SET statement and introduce the BY and MERGE statements and several popular dataset options. Bear in mind that *many* combining techniques are not addressed here because of space constraints.

Methods

Virtually any combination of datasets is possible using SAS. This chapter presents usage rules and syntax for three common techniques.

Concatenation

This is the simplest form of combining. A list of datasets is specified in the SET statement. SAS creates the new dataset by writing every observation from the first dataset, every observation from the second dataset, and so on until all datasets have been read. The result is the "stacking" of datasets in the order in which they were found in the SET statement. The first scenario at the beginning of this chapter illustrates concatenation.

Interleaving

This form of combining closely resembles concatenation. The difference is that you need to use a BY statement. This statement tells SAS to interleave according to the specified variables. If the BY statement specified ID, the new dataset observation order would be all observations with the lowest ID value, followed by all observations with the next lowest ID value, and so on. The second scenario, above, illustrates interleaving.

Matched Merge

Probably the most widely used method for combining datasets, the matched merge enables datasets to be joined side by side, in contrast to the previous methods' stacking. One or more BY variables are specified. The MERGE statement identifies the datasets to be processed. The new dataset is created by linking observations in the MERGE list according to unique levels of the BY variables. For instance, if we merge datasets `demog` and `visit` by `patid`, the new dataset will contain observations with `demog` and `visit` data for `patid` 100, followed by `demog` and `visit` data for `patid` 102, continuing to the highest `patid` value. The third scenario, above, illustrates a matched merge.

Terminology

The dataset-combining features of SAS are easily implemented, but should be used with care. Subtle errors of logic not identified by the SAS syntax checker can lead to unexpected results. Sometimes the errors are obvious, sometimes not. Pay particular attention to Notes and Warnings in the Log, and use complete or partial PRINTs to check your work. Consider these general notes as you work:

- ❑ *Dataset order matters.* The order of the datasets in the SET or MERGE statement dataset lists is significant. In the SET statement, the order determines which dataset is used first. The first, or leftmost, dataset's observations are written to the output dataset first, followed by those from the second dataset. In the MERGE statement, the order determines variable overwriting. If datasets have identically named variables, the variables in the rightmost dataset overwrite those of the dataset to its left. The values are continually being overwritten as observations from new datasets are read in. This can, intentionally or otherwise, cause missing values to replace nonmissing values.

- ❑ *Observation order matters.* The order of the observations is also significant when you are interleaving or performing a matched merge. Datasets must be sorted in the order specified by the BY statement. If they are not in this order, and an error is generated. The SORT procedure is described in Chapter 6.

- ❑ *Data types must match.* If a variable is found in more than one of the datasets being combined, it must be the same data type in *every* dataset. For example, REGION cannot be 1, 2, and 3 (numeric) in one dataset and NE, SE, and MW (character) or '1', '2', and '3' (character) in another. Such a mismatch will cause an error.

- ❑ *Observations "belong" to a dataset.* When SAS is building the new dataset, it keeps track of which SET or MERGE dataset is contributing to the current observation. This information can be captured by using the IN dataset option. See "Dataset Options," below, for more information about IN.

- ❑ *Output dataset size is predictable.* As a way to check your work, recall a few traits of output datasets. These do not consider the impact of the OUTPUT statement, which usually decreases the expected observation count of a dataset. First, datasets created by concatenation or interleaving will be equal to the observation counts in the SET statement datasets. Matched merge output datasets may also be equal to the counts of the individual dataset observations (indicating a complete lack of BY variable matching). In practice, though, they are usually smaller.

  The number of variables is also predictable. The output dataset's number of variables equals the number of unique variable names in all the SET or MERGE datasets. Thus, if two datasets had 10 variables each and all were named differently, the output dataset would have 20 variables. If, however, we were SETting two survey datasets with identical names, we would have 10 variables. Keep in mind that you should be able to explain the Log's Notes about both number of observations and number of variables.

Dataset Options

All of the combining methods described below can utilize various dataset options. One, OBS, was described in Chapter 4. This section describes some others and how they might be used when combining datasets. Their use is demonstrated later in the chapter.

- IN Identifies a numeric variable whose value is 1 if the current observation uses data from the dataset, 0 otherwise. The IN option variables are useful for identifying different combinations of presence and absence. By using the IN option you can, say, specify that the new dataset must consist only of observations that were present in both datasets of a matched merge. Note that the variable specified by the IN option is temporary, existing only for the duration of the DATA step.

  Syntax: `IN=variable_name`

- RENAME Renames variables in the dataset as they are being read in for use when constructing the output dataset's observation. Useful when you want to force variables in different datasets to have similar names. For example, dataset JAN's variable TEMPJAN and dataset SEPT's variable TEMPSEPT may be measuring the same item (average temperature). You would use the RENAME option to standardize the name to, say, TEMP. More than one renaming can take place, as shown below.

  Syntax: `RENAME=(old=new ...)`

Concatenation

Concatenation of datasets requires a simple extension to the SET statement (first described in Chapter 4):

```
SET dataset_name1 ... dataset_namen;
```

Here, `dataset_name1` through `dataset_namen` identify the datasets to be combined. They may be any combination of temporary and permanent (one- and two-level) datasets. Dataset options (described in the preceding section) may be used with any of the datasets.

Remember, the number of observations in the output dataset will equal the sum of the number of observations in `dataset_name1` through `dataset_namen`. The number of variables will range from the number in any dataset (identical dataset structure in all datasets) to the sum of the number of variables in each dataset (each dataset had a different structure). If a variable is in dataset A but not in dataset B, it will be missing for every observation in the concatenated B's (see Exhibit 5.2, below).

Examples

Exhibit 5.1 is the "purest" illustration of concatenated datasets. Each dataset has identical variable names. The output dataset CONCAT is simply dataset JAN stacked on top of dataset JUN.

Exhibit 5.1

Concatenate Two Datasets with Identically Named Variables.

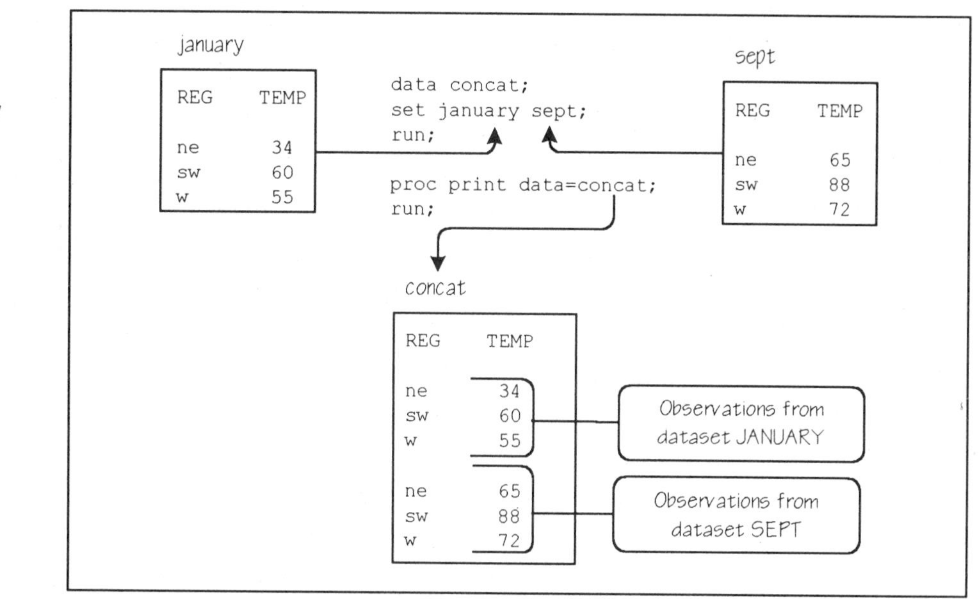

Exhibit 5.2 muddies things slightly. Temperature measurements for January have the variable name `tempjan`; those for September, `tempsept`. The merge creates gaps for variables unknown to the two datasets: `tempsept` was not in dataset **`JANUARY`**, and is therefore missing in all of the January observations. Likewise, `tempjan` was not in **`SEPT`**, and is consistently missing in **`CONCAT`** observations coming from dataset **`SEPT`**.

The variable-naming convention used in Exhibit 5.2 is not unreasonable. It makes sense to convey as much information as possible in a variable name. Exhibit 5.3 shows how these helpful names can be combined when merging in order to eliminate the missing values of Exhibit 5.2. The RENAME option preserves the variable names in datasets `january` and `sept`, renaming them to `temp` during the creation of dataset `concat`. The output dataset contains a single temperature variable (`temp`) instead of two (`tempjan` and `tempsept`).

Exhibit 5.2

Conatenate Two Datasets. Not All Variable Names Are in Both Datasets.

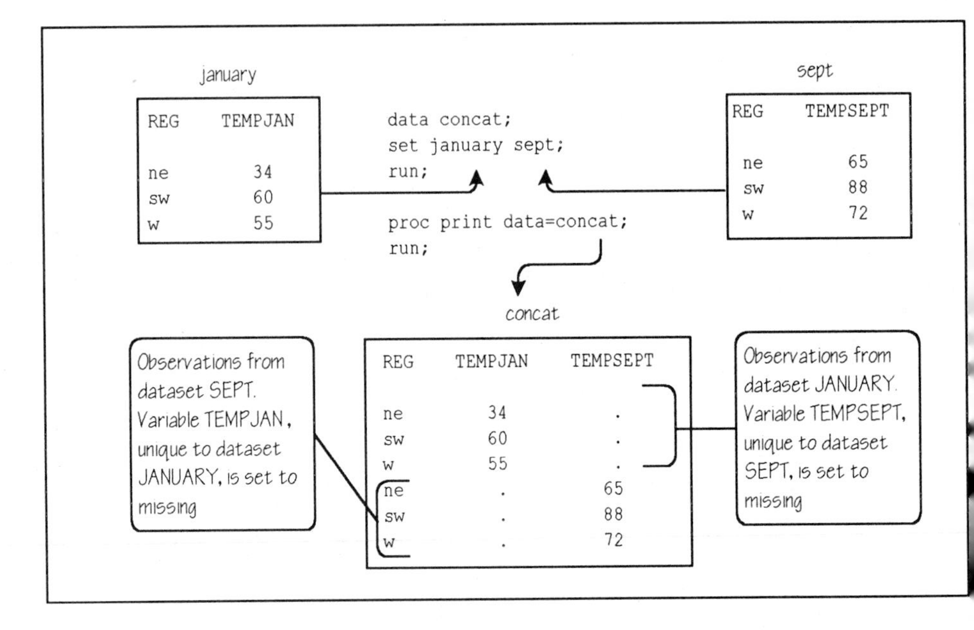

Exhibit 5.3

Use RENAME Option When Concatenating Datasets.

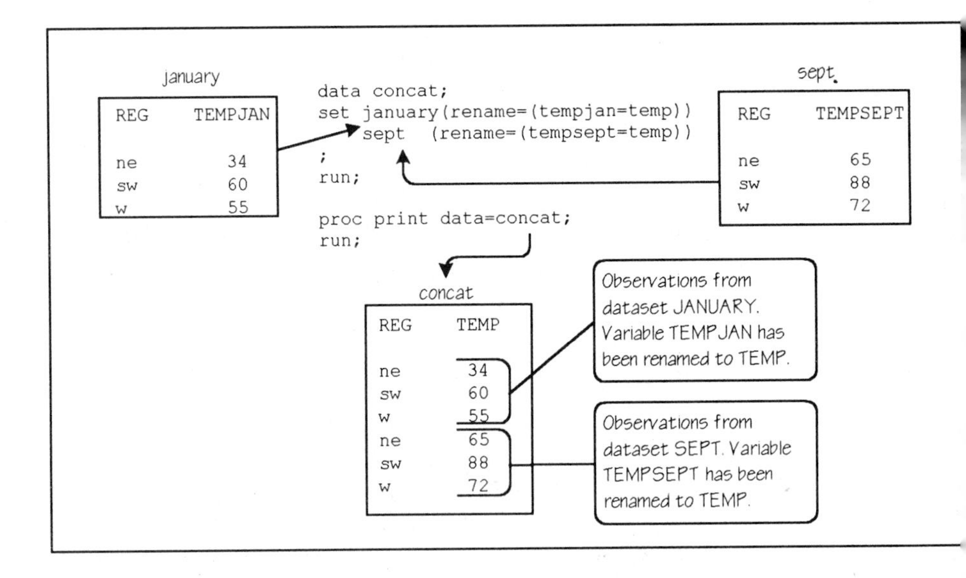

In Exhibit 5.3, we eliminated missing values by using a common variable name. The price we paid is the loss of the variable's origin. It is impossible to tell which observations in Exhibit 5.3 came from `january` and which came from `sept`. This information may be needed to perform analyses stratified by month — a *t* test, for example. The typical solution is to create a variable based on the value of the IN option variable. This is demonstrated in Exhibit 5.4.

Exhibit 5.4

Use Dataset Option IN to Identify the Dataset an Observation Comes From.

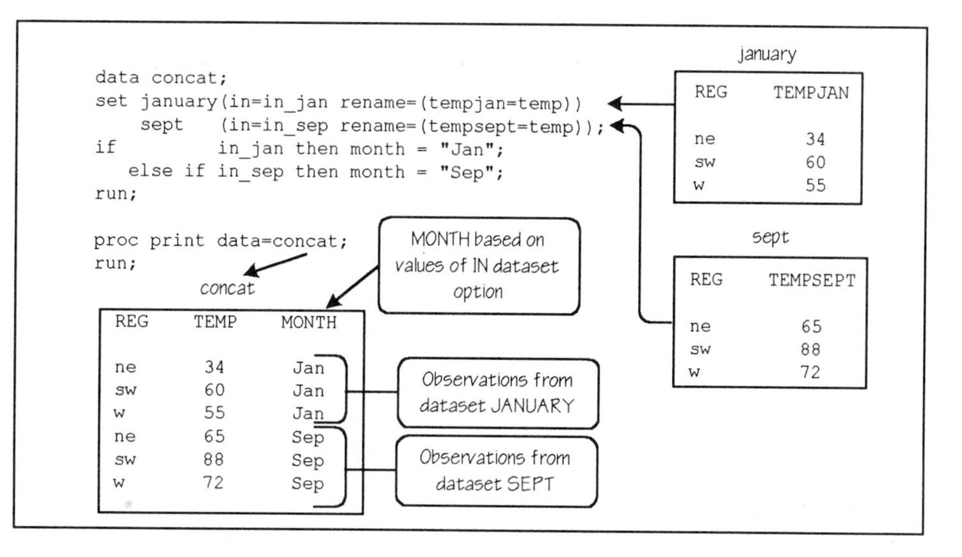
data concat;
set january(in=in_jan rename=(tempjan=temp))
    sept   (in=in_sep rename=(tempsept=temp));
if   in_jan then month = "Jan";
   else if in_sep then month = "Sep";
run;
proc print data=concat;
run;
january
REG TEMPJAN
ne 34
sw 60
w 55
sept
REG TEMPSEPT
ne 65
sw 88
w 72
concat
REG TEMP MONTH
ne 34 Jan
sw 60 Jan
w 55 Jan
ne 65 Sep
sw 88 Sep
w 72 Sep
MONTH based on values of IN dataset option
Observations from dataset JANUARY
Observations from dataset SEPT

Interleaving

Interleaving requires a simple extension to the concatenation syntax described above. Use a BY statement in addition to the SET statement, as shown here:

```
SET dataset_name1 ... dataset_namen;
BY byvar ...;
```

In the above, *dataset_name1* through *dataset_namen* are the datasets to be interleaved and *byvar* is one or more variables indicating the order of interleaving. Each dataset must contain *byvar* and be sorted by each variable in the BY statement. SAS builds the new dataset by taking observations with the lowest value of *byvar* from each dataset, then taking observations with the next lowest value from each dataset, and so on to the end of the data. This means that all observations from the datasets with the same value of *byvar* will be adjacent to each other in the output dataset.

Example

Exhibit 5.5 illustrates the BY statement's impact when it is used with a SET statement. Values of BY variable `id` are scanned in each dataset. Since `round2` has the lower value (19), it contributes an observation first. Then `round1`'s observations with `id` value 20 are used, then `round2`'s observation with `id` value 20. This continues until the last observation, containing **ROUND2**'s value of 31, is used. Notice that `score` is common to both datasets, but `ratng` and `rating` are not. The PRINT procedure highlights what may be a misspelled variable name.

Exhibit 5.5

Interleave Two Datasets.

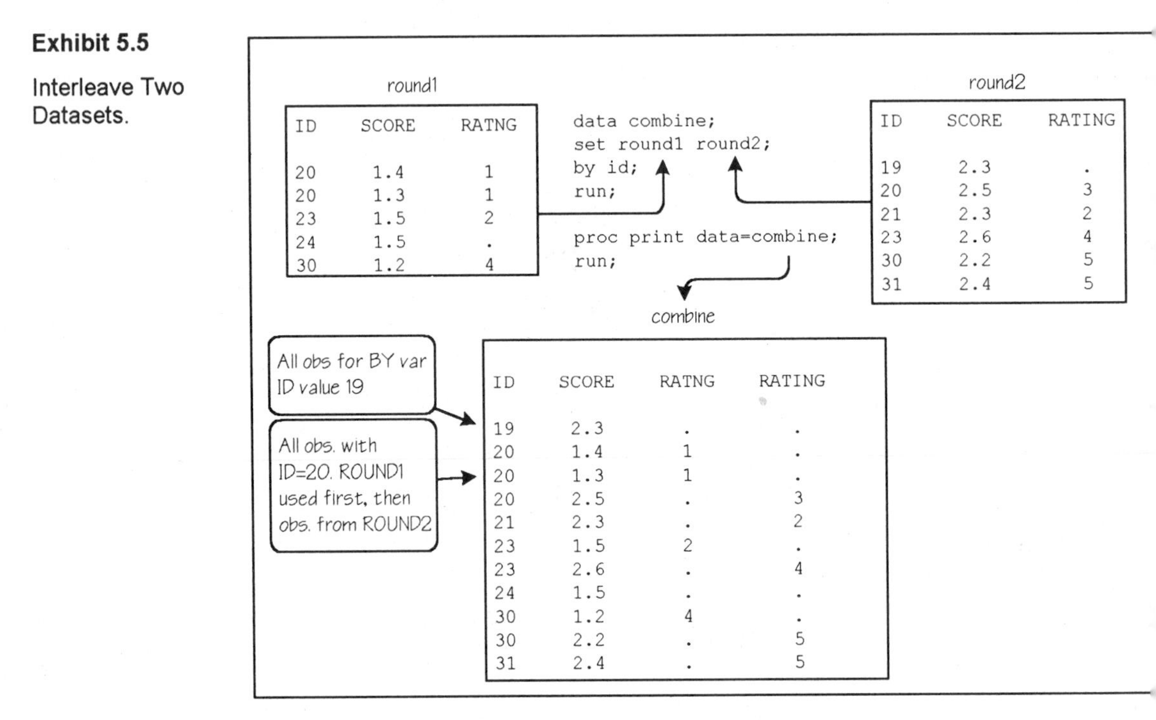

Matched Merge

The matched merge allows combining of two or more datasets by a common variable. Unlike the earlier variations on data combining, the matched merge connects observations from different datasets. Observations with a given BY variable value in one dataset are linked with the corresponding observations in one or more other datasets. That is, observations are *matched* according to BY variable values. The observation in the new dataset contains all the variables of each dataset. This simple and powerful feature allows combining of disparate sources of data. Its use requires a new statement as well as some words of caution.

Syntax and Usage

Use the MERGE and BY statements to perform a matched merge.

```
MERGE dataset_name1 ... dataset_namen;
BY byvar ...;
```

Here, `dataset_name1` through `dataset_namen` identify the datasets to be linked and `byvar` identifies the variables forming the link. They may be any combination of temporary and permanent (one- and two-level) datasets. Dataset options (described earlier in this chapter) may be used with any of the datasets.

A few features of the matched merge should be kept in mind:

- If a variable is found in more than one dataset, the value of the variable in the rightmost dataset will overwrite those to its left. Consider the following statements:

  ```
  DATA both;
  MERGE baseline pass1;
  BY socsec;
  ```

  Suppose variable `income` is in both datasets. If there is a `socsec` value that is common to both MERGE datasets, the value of `pass1` variable `income` will overwrite that coming from `baseline`. Dataset `both` will contain the `income` value from `pass1`. Even if `pass1` had a missing value, it would overwrite a nonmising value in `baseline`. This is demonstrated in Exhibit 5.6.

- The joining does not have to be one-to-one. You could, for example, merge patient demographic data with incident data so that the incident data could be analyzed by categories of demographic variables ("Did pulse elevation after taking the medication vary by age?"). These types of merges are commonly called *many-to-one*. No special coding is necessary to do this, but you do need to be aware of the persistence of variable values in the smaller BY group. This is demonstrated in Exhibit 5.7.

- If a BY variable value has no match in another MERGE dataset, the observation is written to the output dataset with missing values for the variables from the unmatched datasets.

- The IN option may be used to control the contents of the output dataset. You could restrict output to only those observations with a match in both datasets, for example. This is shown below — if an observation was in both `baseline` (IN variable `base` equals 1) and pass1 (IN variable `pass` equals 1), we output to dataset `onlyboth`.

  ```
  DATA onlyboth;
  MERGE baseline(IN=base) pass1(IN=pass);
  BY socsec;
  IF base = 1 AND pass = 1 THEN OUTPUT;
  ```

Examples

Exhibit 5.6 demonstrates many of the features described in the preceding discussion. Two datasets are merged by BY variable `id`. Some values of `id` are common to both datasets; some are not. Examine the output dataset to see how missing values are propagated for the nonmatching observations. Some variables (`id`, `value`) are in both datasets; some (`v_one`, `v_two`) are not. The pattern of IN option variables is also significant. IN variables are used for determining output dataset contents in a later example. Finally, note the overwriting that takes place. The variable `value` is in both datasets. When we have a match on BY variable `id`, values of variable `value` from dataset `two` overwrite those just read in from dataset `one` (for `id` of "a", "." replaces 1.1; for `id` of "b", 2.1 replaces 1.2, and so on).

Exhibit 5.6

Matched Merge Using IN Dataset Option

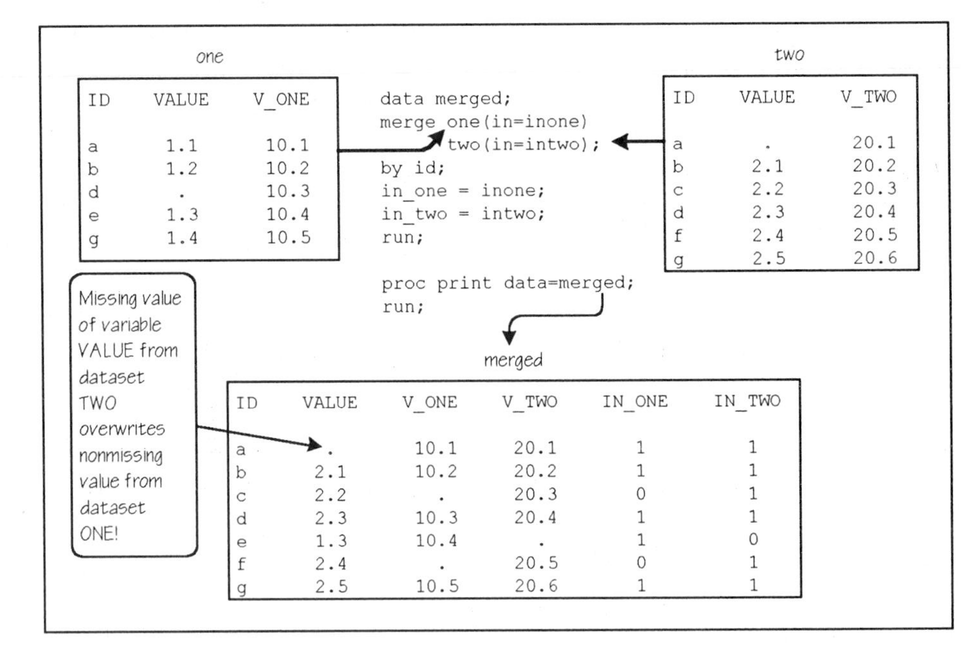

ID	VALUE	V_ONE
a	1.1	10.1
b	1.2	10.2
d	.	10.3
e	1.3	10.4
g	1.4	10.5

```
data merged;
merge one(in=inone)
      two(in=intwo);
by id;
in_one = inone;
in_two = intwo;
run;

proc print data=merged;
run;
```

ID	VALUE	V_TWO
a	.	20.1
b	2.1	20.2
c	2.2	20.3
d	2.3	20.4
f	2.4	20.5
g	2.5	20.6

ID	VALUE	V_ONE	V_TWO	IN_ONE	IN_TWO
a	.	10.1	20.1	1	1
b	2.1	10.2	20.2	1	1
c	2.2	.	20.3	0	1
d	2.3	10.3	20.4	1	1
e	1.3	10.4	.	1	0
f	2.4	.	20.5	0	1
g	2.5	10.5	20.6	1	1

The datasets in Exhibit 5.6 had one observation for each value of the BY variable. That is, in each dataset there was one and only one instance of `id` values "a", "b", and so on through "g". Exhibit 5.7 illustrates SAS's handling of situations where there are multiple occurrences of a BY value in the datasets.

In Exhibit 5.7, pay particular attention to the treatment of "short" datasets: `id` value 4 has two observations in dataset `two` but only one in dataset `one`. *The values of variables unique to* `v1` *stay the same throughout the BY group.* Once SAS runs out of values for `v1`, it simply keeps the last one it had for the current BY group level. This retention of values is also present for `id` values 3 and 5.

When two or more datasets used in a MERGE statement have multiple occurrences of BY variables, a message similar to the following appears in the Log:

```
NOTE: MERGE statement has more than one data set
with repeats of BY values.
```

Exhibit 5.7

Matched Merge with Multiple Observations per BY Group

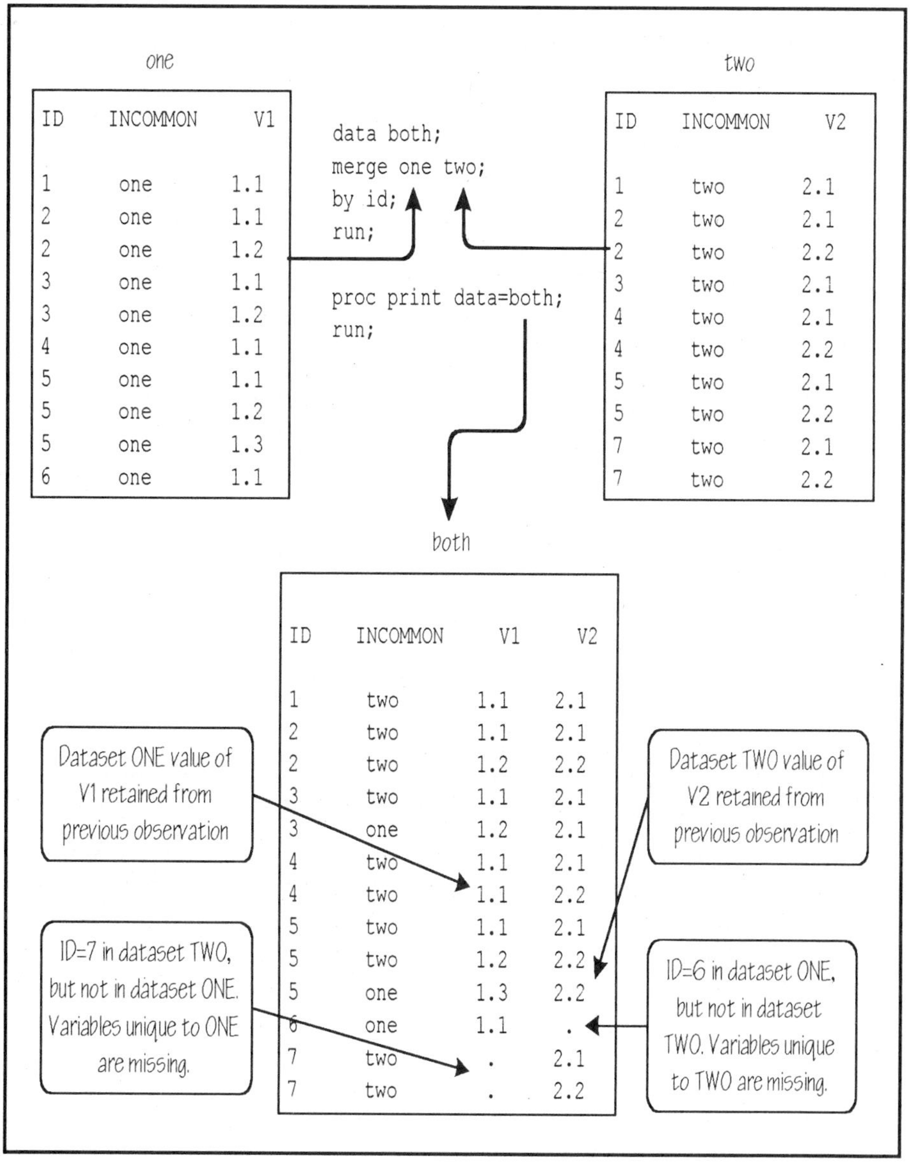

one

ID	INCOMMON	V1
1	one	1.1
2	one	1.1
2	one	1.2
3	one	1.1
3	one	1.2
4	one	1.1
5	one	1.1
5	one	1.2
5	one	1.3
6	one	1.1

```
data both;
merge one two;
by id;
run;

proc print data=both;
run;
```

two

ID	INCOMMON	V2
1	two	2.1
2	two	2.1
2	two	2.2
3	two	2.1
4	two	2.1
4	two	2.2
5	two	2.1
5	two	2.2
7	two	2.1
7	two	2.2

both

ID	INCOMMON	V1	V2
1	two	1.1	2.1
2	two	1.1	2.1
2	two	1.2	2.2
3	two	1.1	2.1
3	one	1.2	2.1
4	two	1.1	2.1
4	two	1.1	2.2
5	two	1.1	2.1
5	two	1.2	2.2
5	one	1.3	2.2
6	one	1.1	.
7	two	.	2.1
7	two	.	2.2

Dataset ONE value of V1 retained from previous observation

Dataset TWO value of V2 retained from previous observation

ID=7 in dataset TWO, but not in dataset ONE. Variables unique to ONE are missing.

ID=6 in dataset ONE, but not in dataset TWO. Variables unique to TWO are missing.

Finally, in Exhibit 5.8, we use IN variables to control the contents of the output dataset. In earlier examples, we did not use an OUTPUT statement. This means that all observations, matched or unmatched on the BY variables, were written to the output dataset. If we wanted to restrict the dataset contents to observations with a match on `id`, we would use the code shown in the exhibit. Other variations of IN variable usage are also possible; for example, output *only* mismatches for error-reporting purposes.

Exhibit 5.8

Use IN option to control contents of the output dataset.

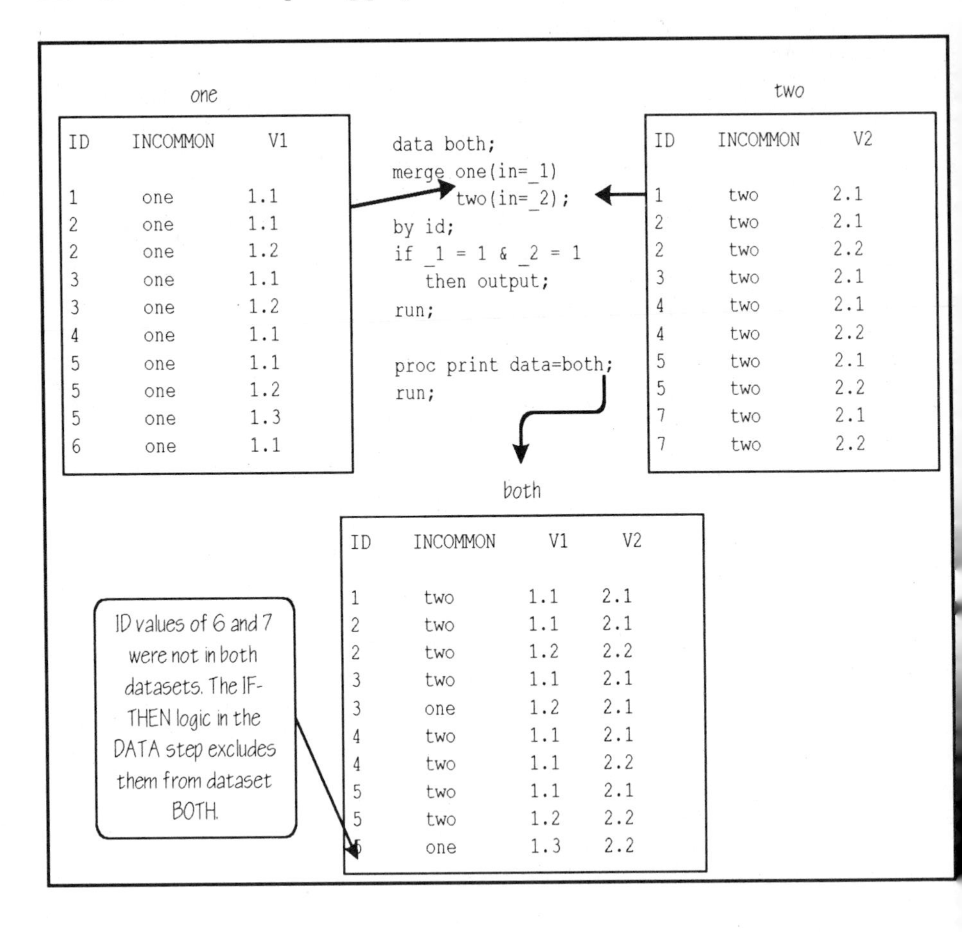

Recap

Syntax Summary

In this chapter, we have introduced the following tools for combining SAS datasets. (See the syntax description in Chapter 1 for an explanation of the notation.)

```
SET|MERGE dataset_name1
    <(IN=var_name RENAME=(old_name=new_name ...)>
    dataset_name2
    <(IN=var_name RENAME=(old_name=new_name ...)>;
<BY <DESCENDING> var_name ...;>
```

Content Review

The SAS System offers many tools for combining datasets. This chapter has discussed the most common applications of a few of those tools. The SET, MERGE, and BY statements enable concatenation, interleaving, and one-to-one merging of two or more SAS datasets. Although the syntax is simple, the results can be puzzling if you do not understand SAS's logic. The chapter presented numerous examples to illustrate the impact of different strategies on the structure of the new, combined dataset.

You now have a small but powerful collection of data management tools at your disposal. Chapter 6 presents an introduction to using procedures. It describes statements common to all procedures, then outlines the use of a few essential utility procedures.

Introduction to Procedures

In the earlier chapters, we reviewed the relatively few SAS statements that identify the source and the layout of your data. We explained how to store the SAS dataset so it could be used in later programs and how to perform calculations, and we outlined the rules for combining two or more SAS datasets.

Fine. But now that you've created the dataset, what can you do with it? Examine and analyze it, using procedures. Recall the discussion in Chapter 2 of how SAS organizes units of work. Generally, DATA steps create and modify SAS datasets, while procedures (PROCs) analyze the data. Recall too that PROCs operate only on SAS datasets.

This chapter discusses some of the more useful, albeit nonstatistical, procedures. PRINT displays the values of some or all of the variables in the dataset. SORT rearranges the dataset by one or more variables. CONTENTS displays useful information about the dataset — items such as the number of observations and the variable names and types.

The last section of this chapter describes statements that can be used with virtually all the procedures discussed in this book. TITLE and FOOTNOTE provide the means to annotate PROC output. FORMAT improves the appearance of variables in output. LABEL enhances the description of variable names. BY analyzes a dataset by subgroups. Finally, WHERE enables you to use only those observations whose variables meet certain logical conditions.

Statements Used with Most Procedures

The procedures described in this book are quite useful by themselves. Sometimes, though, you need to supplement their output. It's helpful to control the display of the data, provide a description of the variables, add titles and footnotes, and so on. This section describes the syntax, and the use, of some of the more useful statements that can be used with most procedures.

Clarifying Content: TITLE and FOOTNOTE

The default title line at the top of a page is usually "The SAS System." The default footnote is a blank line. You can supply more meaningful titles and footnotes by using TITLE and FOOTNOTE statements. The syntax is

```
TITLEx "text";
TITLEx 'text';
FOOTNOTEx "text";
FOOTNOTEx 'text';
```

Here, *x* indicates the first through tenth titles and footnotes. `text` can be anything that improves the readability and usefulness of the listing. It is often helpful to use a footnote to identify the name of the program that generated the listing. Titles and footnotes remain in effect from procedure to procedure.

There are only two rules to keep in mind when using titles and footnotes. First, if the text contains a quotation mark ("), represent it with a pair of quotes ("") in text enclosed by quotes or, better, enclose the entire text between apostrophes ('). Second, replacing a TITLE or a FOOTNOTE removes higher-numbered titles and footnotes. For example, consider the following portion of a SAS program. (The PRINT procedure is described later in this chapter. Here and elsewhere in this section, just focus on the use of the TITLE and FOOTNOTE statements.)

```
PROC PRINT DATA=survey.pass1;
TITLE1 "STAT 100: Survey Data - Pilot";
TITLE2 "Administered to Local Town Planners";
FOOTNOTE "Program is in file C:\stat100\asgn01.sas";
RUN;
```

Two title lines will be used with the PRINT output. Now suppose we enter the following statements:

```
PROC PRINT DATA=survey.final;
TITLE1 "Final Version of Survey";
RUN;
```

Because `TITLE1` was entered, higher-numbered titles (in this case, title 2) are lost. The second PRINT will have only a single title line.

Processing Data in Groups: BY

Most SAS procedures will, by default, process all observations from the dataset at once. Some (REGRESSION, FREQ, and others) can create and analyze subgroups of the data. Others, such as CONTENTS, do not have this capability. The BY statement enables you to run virtually all procedures for each level of the BY variables. If this BY-group processing is not appropriate for a procedure, the BY statement is ignored. The only requirement for BY-group processing is that the dataset be arranged in the order of the variables in the BY statement (see the SORT procedure, described later in this chapter).

When you use a BY statement with a procedure, your processing is a bit different than usual. First, the procedure is repeated for each combination of the BY statement variables. Normally, all observations are treated as a single group. Suppose your dataset had five levels of variable `grade` in each of seven `regions`. Suppose further that your program had the following statements:

```
PROC PRINT DATA=master.roster;
BY region grade;
```

The PRINT procedure would separate the output into 35 distinct groups, one for each combination of the seven regions and five grades.

The second aspect of BY-group processing is the appearance of the output. The example above has 35 distinct levels of BY-group variables. The output, in turn, needs to indicate which group is being displayed. This varies considerably from procedure to procedure. PRINT, for example, displays a row of dashes across the output and identifies the variables in the group and their values. Other procedures make it equally obvious that analysis for a new group is beginning. For now, just remember that the output's appearance will be different and that this is normal.

A final point about BY-group processing: The procedure treats the subgroups separately. Thus, there is usually no "roll–up" or aggregation to the dataset level. If you ran a regression with a BY variable of `race`, the analysis would be repeated for each level of `race`, *but not for the entire dataset*. That analysis requires re-entry of the regression statements, this time omitting the BY statement.

Controlling Appearance: FORMAT and LABEL Revisited

Chapter 4 described the benefits and syntax of the FORMAT and LABEL statements. These provide ways to better display data and to expand a variable name into a longer, more meaningful description. In Chapter 4, however, the context was that of individual procedures. That is, the FORMAT or LABEL statement was in effect only for the duration of the procedure.

Both FORMAT and LABEL may be used as part of a DATA step. The syntax is *exactly* the same as described in Chapter 4. The only difference is scope. When used in a DATA step, the formats and labels are attached to the dataset. Whenever the dataset is used, the formats and labels will be available. This saves you the trouble of duplicating or re-entering the same FORMAT or LABEL statement for multiple procedures. If necessary, you can override a format specification or label for a specific procedure. This will not affect what's stored in the dataset. Consider the following example:

```
DATA demog;
/* INFILE, other statements go here ... */
FORMAT baseline 7. st_grade z2.;
LABEL baseline = "Starting salary";
RUN;

PROC PRINT DATA=demog;
FORMAT baseline dollar10.;
RUN;

PROC FREQ DATA=demog;
LABEL st_grade = "Starting Pay Grade";
RUN;
```

The PRINT procedure's FORMAT overrides the format specification stored in the dataset `demog`. The `7.` format is still the format stored in the dataset. Likewise, in the FREQ procedure, we attach a label to `st_grade` for use with the PROC, but in the dataset we still have the original (blank) label.

Selecting Observations: WHERE

The last topic in this section is observation selection. Most procedures will, by default, use all appropriate observations. PRINT will display every observation in the dataset. REGRESSION will use observations with nonmissing numeric values.

You can limit or add to the inclusion of observations with the WHERE statement. WHERE says, in effect, "The procedure should use only those observations that meet certain criteria." You may want to create a correlation matrix for companies with a particular industry code, or print observations for males with low test scores.

WHERE's syntax is straightforward:

```
WHERE condition;
```

Here, *condition* is a logical expression using constants and/or one or more variables from the dataset being analyzed. Its syntax is exactly the same as that used in IF-THEN-ELSE statements (see Chapter 4). If *condition* is true for an observation, the procedure will use that observation from the analysis dataset. Consider the following examples:

```
PROC PRINT DATA=surv.wavel;
WHERE area = 10 & gender = 2;
```

PRINT will display only those observations from `surv.wavel` for which `area` equals 10 and `gender` equals 2.

```
PROC PRINT DATA=test;
VAR calcl rate duration;
WHERE calcl = .;
```

PRINT will display observations from `test` where `calcl` is missing. If `rate` and `duration` were used to calculate `calcl`, we could inspect the output to see if there was a pattern of missing values or other problems with the data.

Finally, notice a subtle but important aspect of WHERE's usage. It replaces a preliminary DATA step, enabling you to filter "on the fly" if the variables used in the filtering already exist in the dataset. Look at the following program fragment:

```
DATA subset;
SET datain.survey;
IF score > 40 THEN OUTPUT;
RUN;

PROC PRINT DATA=subset;
RUN;
```

The DATA step creates `subset` by reading every observation from dataset `datain.survey` and selecting only those observations where `score` exceeds 40. This could use considerable computer resources (and waste your time as well) if dataset `datain.survey` is large. A more effective way to produce the same result is

```
PROC PRINT DATA=datain.survey;
WHERE score > 40;
RUN;
```

Identifying Groups of Variables: Variable Lists

The procedures discussed throughout this book may require that you specify a list of variables to be processed. In the PRINT procedure, for example, you specify a list of variables to be displayed. Suppose you want to see variables `id`, `gender`, and `income1` through `income10`. One way would be to enter the following list:

```
id gender income1 income2 income3 income4 income5
income6 income7 income8 income9 income10
```

Think of how tedious this coding would become if there were 50 income variables! Fortunately, SAS provides a shortcut notation for variable lists. Its general form is

```
VARstart-VARend
```

Here, `VAR` is the common portion of all the variables in the list (e.g., `income`); *`start`* and *`end`* are the beginning and ending suffixes, respectively. Now we can rewrite our example list as

```
id gender income1-income10
```

There are only a few restrictions on the use of variable lists:

- The *`start`* value must be a number smaller than the *`end`* value. `income1-income10` is valid, but `income10-income1` is not.
- All the variables in the list must be in the dataset you are using. If not, SAS will print an error message and terminate the procedure. If the dataset contained only variables `income1` through `income7`, `income9`, and `income10`, using the variable list `income1-income10` would create an error.

Displaying the Dataset: PRINT

One of the most common activities in data analysis is simply displaying your data. This is especially important during the early stages of research, since you are still getting a feel for the way the data are represented, the number and pattern of missing values, and the like. The PRINT procedure provides a simple way to display all or part of your dataset. It can sum numeric variables and count observations. Some of its options give you control over the display. These are all addressed below.

Example

Exhibit 6.1 uses the airport data from earlier chapters to illustrate the output produced by PRINT when we take all display and calculation defaults. See the exhibit annotation for notes about the output's components.

Exhibit 6.1

Using PRINT Procedure Defaults

Program

```
LIBNAME sample 'dataset location';
PROC PRINT DATA=sample.airport;
RUN;
```

Exhibit 6.1

Using PRINT Procedure Defaults (continued)

Output

OBS	AIRPORT	CITY	ST	SCH_DEP	PERF_DEP
1	HARTSFIELD INTL	ATLANTA	GA	285,693	288,803
2	BALTO/WASH INTL	BALTIMORE	MD	73,300	74,048
3	LOGAN INTL	BOSTON	MA	114,153	115,524
4	DOUGLAS MUNI	CHARLOTTE	NC	120,210	121,798
5	MIDWAY	CHICAGO	IL	64,465	66,389
6	O'HARE INTL	CHICAGO	IL	322,430	332,338
7	DALLAS/FT WORTH INTL	DALLAS/FT WORTH	TX	266,737	269,665
8	LOVE FIELD	DALLAS/FT WORTH	TX	39,481	40,196
9	STAPLETON INTL	DENVER	CO	154,067	156,293
10	DETROIT CITY	DETROIT	MI	6,828	7,162
11	WAYNE COUNTY	DETROIT	MI	134,929	137,565
12	WILLOW RUN	DETROIT	MI	4,241	4,024
13	HONOLULU INTL	HONOLULU	HI	92,659	96,780
14	INTERCONTINENTAL	HOUSTON	TX	104,249	105,330
15	HOBBY	HOUSTON	TX	61,387	62,582
16	ELLINGTON FIELD	HOUSTON	TX	1,188	1,253
17	MC CARRAN INTL	LAS VEGAS	NV	92,196	92,072
18	HOLLYWOOD-BURBANK	LOS ANGELES	CA	30,444	30,968
19	LONG BEACH	LOS ANGELES	CA	14,443	14,712
20	LOS ANGELES INTL	LOS ANGELES	CA	213,302	215,740

OBS	PASSNGER	FREIGHT	MAIL	SCH2PERF	TOT_TONS	PASS_DEP
1	22,665,665	165,669	93,039	98.9	258708.24	78.4814
2	4,420,425	18,042	19,723	99.0	37764.45	59.6968
3	9,549,585	127,815	29,786	98.8	157600.81	82.6632
4	7,076,954	36,243	15,399	98.7	51642.30	58.1040
5	3,547,040	4,495	4,486	97.1	8980.36	53.4281
6	25,636,383	300,464	140,359	97.0	440823.18	77.1395
7	22,899,267	142,661	86,707	98.9	229367.71	84.9175
8	2,882,836	2,217	243	98.2	2459.57	71.7195
9	11,961,839	67,346	38,044	98.6	105389.48	76.5347
10	362,655	258	0	95.3	258.08	50.6360
11	9,903,078	42,831	32,430	98.1	75260.98	71.9884
12	35	33,858	1,249	105	35107.26	0.0087
13	9,002,217	139,497	19,951	95.7	159447.94	93.0173
14	7,543,899	62,425	21,074	99.0	83499.21	71.6216
15	3,972,327	3,788	790	98.1	4578.26	63.4740
16	18,967	199	1	94.8	200.91	15.1373
17	7,796,218	11,289	13,132	100	24420.85	84.6752
18	1,698,739	6,415	1,673	98.3	8087.88	54.8547
19	692,995	7,838	930	98.2	8767.94	47.1041
20	18,438,056	352,824	71,589	98.9	424412.34	85.4642

Identifying the Dataset

The only item PRINT needs is the name of the dataset to be displayed. This is done using the DATA parameter. Its form is

```
PROC PRINT DATA=dataset_name;
```

where *dataset_name* is a one- or two-level dataset name.

Sometimes you may want to test a procedure on just a portion of the dataset's observations. This is often the case when you are using a new procedure. To control the number of observations displayed, use the OBS option following the dataset name:

```
PROC PRINT DATA=dataset_name(OBS=n_obs);
```

In the above, `n_obs` indicates that the first `n_obs` observations from `dataset_name` should be used. If `n_obs` exceeds the number of observations in the dataset, no message is produced and no error is generated. Once you are satisfied that the procedure is performing as you intended, simply remove the OBS option. This restores the procedure to its default behavior — using all observations in the dataset.

Identifying and Ordering Variables

By default, PRINT displays all variables in the dataset in the order in which they are stored in the dataset. To restrict the variables or to arrange them differently, use the VAR statement:

```
VAR var_names;
```

var_names can be a list containing any number of character or numeric variables in the dataset. Some examples follow:

```
VAR id soc_sec;
VAR age race gender party vote1-vote5;
```

Another default activity shown in Exhibit 6.1 is the identification of each observation by a number (in the column headed `OBS`). You can override this default by using the ID statement:

```
ID var ...;
```

Here, *var* indicates one or more variables that will replace the OBS identifier. Commonly used ID variables are key fields such as social security number, survey number, patient and visit numbers, and other variables that help to uniquely identify an observation.

The ID statement is especially useful when there are many variables in a dataset and PRINT cannot fit all the variables across the page. In this case, it will break up the listing into two or more rows per observation. The ID statement's variable(s) begin each line of an observation, thus making it easier to identify the lines that belong to an observation.

Requesting Counts and Sums

PRINT can perform some basic counting and addition. The N option in the PRINT statement requests a count of the observations displayed. The value is printed at the end of the listing. To request N, use a PROC statement such as

```
PROC PRINT DATA=dataset_name N;
```

Add variables by using the SUM statement. The SUM statement resembles VAR, described earlier:

```
SUM var ...;
```

If you use a BY statement and a SUM statement, PROC PRINT will create group subtotals for each variable in the SUM statement list. Only numeric variables may be specified in the variable list. Entering character variables creates an error and stops program processing.

Aesthetics

The default PRINT output is usually not used for formal presentations. There are, though, a few simple options in the PROC statement that can improve the display's appearance. By default, PRINT determines the best layout on a page-by-page basis. This can result in a tightly packed page, followed by a relatively sparse one, depending on the data values. UNIFORM ensures that the layout of the listing is the same from page to page.

Another PRINT default is using the variable name as the column header in the display. The SPLIT option tells PRINT to use variable labels for the headers. If a label is not found, the variable name is used instead. The syntax is

```
SPLIT="character"
```

`character` is the single character in the label that determines line breaks. This allows you to control the appearance of the column header.

The use of these options is straightforward:

```
PROC PRINT DATA=dataset_name UNIFORM SPLIT=" ";
```

PRINT will use the same layout on each page and will use variable labels instead of variable names as the column headers.

Example

Exhibit 6.2 uses some of the options described above. We sort the `airport` dataset, writing to a temporary dataset, `temp`. In the PROC statement, we use variable labels as column headers (SPLIT option) and request a count of the observations displayed (N option). We use several statements associated with the procedure. The ID statement specifies the variable `airport` as the leftmost identifier in the display. The VAR statement restricts the number of variables to be printed. It also controls the order in which they are displayed. The BY statement requests that PRINT separate the listing into groups of observations with the same value of `st`. SUM requests adding up the values of variable `passnger`, and FORMAT and LABEL are added to improve the presentation of some of the variables used in the procedure.

Exhibit 6.2

Using Aesthetic and Counting Options in the PRINT Procedure

Program

```
LIBNAME sasdata 'c:\book\data\sas';

PROC SORT DATA=sasdata.airport OUT=temp;
BY st;
RUN;

PROC PRINT DATA=temp(OBS=20) N SPLIT=" ";
ID airport;
VAR st passnger pass_dep;
BY st;
SUM passnger;
FORMAT passnger comma10. pass_dep 3.;
LABEL passnger = "Enplaned passengers"
      pass_dep = "Passengers per departure" ;
TITLE "Airport Passenger Data";
RUN;
```

Output

```
                         Airport Passenger Data

---------------------------------- State=AK -------------------

                                                    Passengers
                                    Enplaned           per
               AIRPORT             passengers      departure

               ANCHORAGE INTL       1,362,282          39
               FAIRBANKS INTL         233,809          38
                                   ----------
                    ST              1,596,091

                                   N = 2

---------------------------------- State=AL -------------------

                                                    Passengers
                                    Enplaned           per
               AIRPORT             passengers      departure
```

Exhibit 6.2

Using Aesthetic and Counting Options in the PRINT Procedure (continued)

```
                BIRMINGHAM MUNI       1,001,983          50
                MADISON COUNTY          381,668          38
                BATES FIELD             380,798          39
                                     ----------
                      ST              1,764,449

                                      N = 3

--------------------------------- State=AR ------------------

                                                    Passengers
                                  Enplaned             per
                      AIRPORT    passengers         departure

                   ADAMS FIELD      950,540            62

                                    N = 1

--------------------------------- State=AZ ------------------

                                                    Passengers
                                  Enplaned             per
                AIRPORT          passengers         departure

                SKY HARBOR INTL  10,727,494            72
                TUCSON INTL       1,263,509            62
                                 ----------
                      ST         11,991,003

                                  N =8
Airport Passenger Data
--------------------------------- State=CA ------------------

                                                    Passengers
                                      Enplaned         per
              AIRPORT                passengers     departure

              HOLLYWOOD-BURBANK       1,698,739        55
              LONG BEACH                692,995        47
              LOS ANGELES INTL       18,438,056        85
              ORANGE COUNTY           2,203,700        58
              SAN DIEGO INTL          5,260,907        74
              BUCHANAN FIELD             49,532        37
              OAKLAND METRO INTL      2,670,788        58
              SAN FRANCISCO INTL     13,474,929        72
              ONTARIO INTL            2,640,734        64
              SACRAMENTO METRO        1,737,096        37
              SAN JOSE MUNI           3,128,393        62
              FRESNO AIR TERMINAL       393,442        19
                                     ----------
                    ST               52,389,311
                                     ==========
                                     68,691,394
                                   N = 12
Total N = 20
```

Rearranging the Dataset: SORT

Observations are often found in a dataset in a somewhat random order. Survey data may be entered in the order in which they are returned in the mail. Personnel data may be supplied in Social Security Number order rather than by department. In many situations, the order is irrelevant: we will see in later chapters that some PROCs organize unordered data for you. Sometimes, however, you need to impose order, either because it is easier to use listings or because you want to analyze data by groups and the procedure cannot create the groups for you. The SORT procedure performs this dataset arranging.

Specifying the Datasets

SORT needs to know the names of two datasets. The first is the original, unsorted dataset. The second is the dataset holding the rearranged data. These datasets are specified by the DATA and OUT options in the SORT statement:

```
PROC SORT DATA=original_dataset OUT=new_dataset;
```

Specifying Observation Order: The BY Statement

The other element in the sorting procedure is identifying one or more sort *keys*. These are the variables by which the new dataset will be arranged. Identify them by using the BY statement:

```
BY by_var1 <by_var2 ... > ;
```

Here, *by_var1, by_var2* and other sort keys are any SAS variable. Variables in the BY statement are commonly referred to as BY variables. The output dataset's observations will be organized in *by_var* order, with observations containing the lowest values and earliest alphabetic characters coming first. If more than one BY variable is specified, the rightmost variable will be sorted within levels of variable(s) to its left. Suppose a dataset contains the following observations:

REGION	DPTMENT
a1	9
a1	12
a1	8
b2	10
b1	1
b1	3

If we sort using

```
PROC SORT DATA=dept OUT=sortdep;
BY region dptment;
```

the output dataset looks like this:

REGION	DPTMENT
a1	8
a1	9
a1	12
b1	1
b1	3
b2	10

Note that in some circumstances SAS can determine that the input dataset is already correctly sorted. When this occurs, SAS will simply copy the original to the new dataset. This feature can save large amounts of computer resources.

Sorting Options

SORT produces no printed output, so aesthetics and options are not relevant. One of the few options available is NODUPKEY. When specified in the PROC statement, it eliminates multiple occurrences of observations with duplicate values of BY variables. Another PROC option places more stringent requirements on eliminated observations: NODUPREC eliminates observations with duplicate values in *all* variables, not just BY-variables. When either NODUPKEY or NODUPREC is used, SORT will display a count of the deleted observations in the Log.

By default, SAS arranges observations in ascending, or low-to-high, order. For example, lower-valued state codes precede higher ones, and lower salaries precede higher ones. Sometimes it is helpful to have the highest values come first. If you wanted to report census tracts with the highest education levels, you would need to sort in descending education order. The DESCENDING option in the BY statement does this:

```
BY DESCENDING var ...;
```

This syntax tells SAS to sort the next variable in the list in reverse alphabetic, or high-to-low numeric, order. The BY statement can have a mix of ascending and descending variables, as shown in the following examples:

```
BY region DESCENDING salary;
BY DESCENDING taxrt state DESCENDING pop;
```

Notice that SAS can distinguish the keyword `DESCENDING` from variables in the list because it is too long for a legal SAS variable name (10 characters instead of 8).

Getting Information About the Dataset: CONTENTS

Unlike input, or "raw," datasets, SAS datasets cannot be viewed with text editors or file viewers. The PRINT procedure takes care of this problem up to a point, but sometimes more information is needed. For example, you may need to know the variables' data types, labels, and display formats. Other, more general, information about a dataset may also be needed. Such information includes the number of variables in the dataset, the dates and times the dataset was created and was last updated, and so on. This and other information is displayed by the CONTENTS procedure.

Identifying the Datasets

Since more than one SAS dataset can be stored in a library, you have a bit more flexibility than usual when using the DATA option in the PROC statement. The possible formats are

```
PROC CONTENTS DATA=<libref.>dataset_name;
PROC CONTENTS DATA=<libref.>_ALL_;
```

When you use the first form of the DATA option, CONTENTS displays information about a particular dataset. The special dataset name _ALL_ requests information for every dataset in the location pointed to by the libref.

Controlling the Amount of Output

Sometimes you need only CONTENTS to give you the names of the variables in the dataset. Use the SHORT option in the PROC statement in this situation:

```
PROC CONTENTS DATA=dataset_name SHORT;
```

Example

CONTENTS reports on SAS datasets, and SAS datasets themselves vary from system to system. Exhibit 6.3 presents output from SAS Version 6.10, running under Windows 3.11. Keep in mind that your output from CONTENTS may appear different if you're using a non-Windows operating system.

Exhibit 6.3

Default CONTENTS Output from SAS 6.10 Running Under Windows 3.11

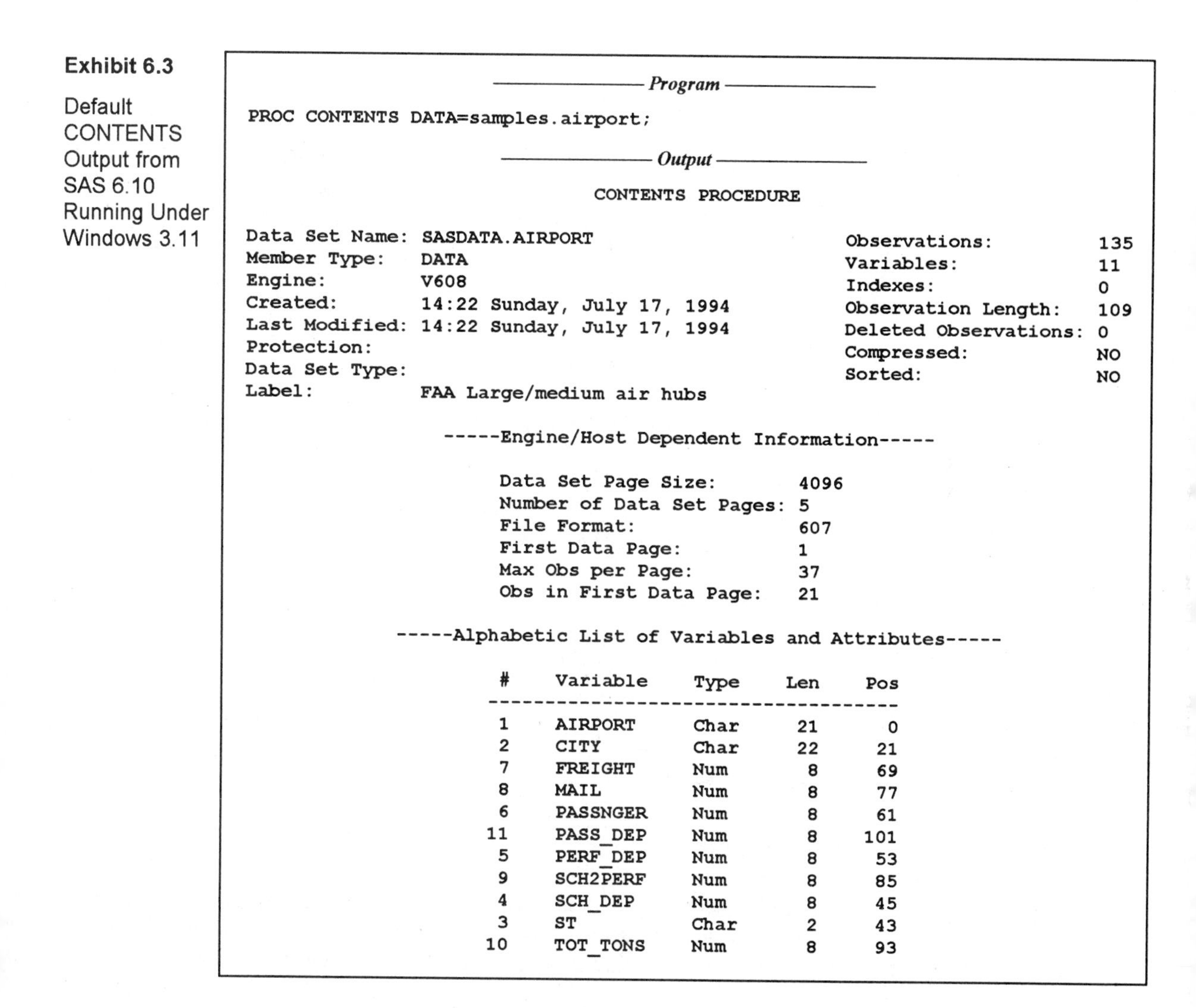

Program

```
PROC CONTENTS DATA=samples.airport;
```

Output

```
                         CONTENTS PROCEDURE

Data Set Name: SASDATA.AIRPORT                       Observations:          135
Member Type:   DATA                                  Variables:             11
Engine:        V608                                  Indexes:               0
Created:       14:22 Sunday, July 17, 1994           Observation Length:    109
Last Modified: 14:22 Sunday, July 17, 1994           Deleted Observations:  0
Protection:                                          Compressed:            NO
Data Set Type:                                       Sorted:                NO
Label:         FAA Large/medium air hubs

                -----Engine/Host Dependent Information-----

                    Data Set Page Size:          4096
                    Number of Data Set Pages:    5
                    File Format:                 607
                    First Data Page:             1
                    Max Obs per Page:            37
                    Obs in First Data Page:      21

            -----Alphabetic List of Variables and Attributes-----

                    #    Variable    Type    Len    Pos
                    -----------------------------------
                    1    AIRPORT     Char     21      0
                    2    CITY        Char     22     21
                    7    FREIGHT     Num       8     69
                    8    MAIL        Num       8     77
                    6    PASSNGER    Num       8     61
                   11    PASS_DEP    Num       8    101
                    5    PERF_DEP    Num       8     53
                    9    SCH2PERF    Num       8     85
                    4    SCH_DEP     Num       8     45
                    3    ST          Char      2     43
                   10    TOT_TONS    Num       8     93
```

Recap

Syntax Summary

This chapter has introduced you to SAS procedures. Some statements may be used with virtually all procedures. (See the syntax description in Chapter 1 for an explanation of the notation.)

```
TITLE<1|2|...|10> "title_text";
FOOTNOTE<1|2|...|10> "footnote_text";
WHERE condition;
```

The SAS System provides a "shorthand" reference to a list of consecutively numbered variables:

```
varstart-varend
```

You may restrict the number of observations to use from a dataset:

```
DATA=dataset_name(OBS=n_obs);
```

The following utility procedures were discussed:

```
PROC PRINT DATA=dataset_name <N> <UNIFORM>
           <SPLIT="character">;
<VAR var_name1 <var_name2 ...> ;>
<ID var ...;>
<SUM numeric_var1 <numeric_var2 ...> ;>

PROC CONTENTS DATA=<dataset_name|_ALL_> <SHORT>;

PROC SORT DATA=original_dataset OUT=sorted_dataset
          <NODUPREC|NODUPKEY>;
BY <DESCENDING> var_name1 <var_name2 ...> ... ;
```

Content Review

Earlier chapters focused on SAS fundamentals and on how to build SAS datasets. In this chapter, we began to use the datasets in procedures, or PROCs. In each of the three PROCs discussed, we saw how relatively few statements can accomplish a good deal of useful work. The PROCs allow you to rearrange, display, and examine the structure of the datasets.

In addition to the procedures themselves, we introduced statements that can be used with virtually any procedure. These statements enable you to add titles and footnotes to output, add labels and formats to variables, process a dataset in subgroups, and filter out unwanted observations from an analysis.

The next chapter begins the statistical portion of the book by examining the range of statistical procedures available in the SAS System.

A Statistical Roadmap for the SAS System

As we mentioned in Chapter 1, two objectives of this book are to give you enough of an introduction to the SAS language to enable you to

- Read and manipulate research data to prepare it for statistical analysis.
- Perform analyses using statistics commonly employed by researchers in the health, management, behavioral, and social sciences.

We have worked on the first objective in the preceding chapters. We begin to work on the second in the following chapters. We also want to provide you with a roadmap for the wide variety of statistical procedures available in the SAS System. In this chapter, we'll give you an overview of the SAS System's considerable statistical capabilities. This overview covers intermediate and advanced statistical methods as well as elementary ones. If you're a statistical novice, you may want to skip directly to Chapter 8. Or you may want to read this chapter just to get an idea of what SAS offers. Don't worry about discussion of statistical techniques you're not familiar with yet.

What's in a PROC Name?

The SAS System implements extremely rich and sophisticated statistical techniques in its many statistical PROCs. The names of some PROCs are very clearly related to statistical methods. You don't have to be a trained statistician to know that PROC MEANS provides means of variables or that PROC REG is the basic PROC for regression. However, not every statistical technique is implemented in a PROC with an obvious name. For example, although box plots of the frequency distribution of variables often used for exploratory data analysis are available in PROC UNIVARIATE, there is no PROC BOXPLOT or PROC EDA.

Another source of confusion is that it may be possible to use several PROCs for an analytical technique. However, it may be that only one PROC can perform the analysis using the correct assumptions for a particular research situation or is the most convenient for a particular purpose. Both PROC ANOVA and PROC GLM may be used for the analysis of variance. However, PROC ANOVA is appropriate only for situations with equal, proportional, or balanced numbers of observations in each treatment group — PROC GLM may be employed for situations where those numbers differ markedly.

This lack of straightforward correspondence between statistical technique name and PROC name often frustrates beginning users of the SAS System. The purpose of this chapter is to provide a roadmap to acquaint you with the rich diversity of statistical techniques that can be performed with SAS. It also helps you find the appropriate PROC for a particular statistical purpose. It does not cover all the statistical techniques implemented in the SAS System, or all the SAS procedures that could be used for a particular technique. We do try to point you toward the most straightforward and generally useful PROCs for those techniques most commonly used in research in the health, management, and social sciences.

A Warning

Please remember that we cannot attempt to teach the theory, assumptions, interpretational nuances, and caveats necessary for intelligent use of the statistical techniques touched on in this chapter. Bear in mind that SAS statistical PROCs are extremely powerful and can produce very sophisticated analyses of data in a fraction of a second. They usually do not, however, check on the level of measurement of the numeric SAS variables used in them. Nor do they check on other assumptions underlying the descriptive and inferential statistics they produce. It is very easy to use SAS to calculate the mean of a nominally scaled variable or to calculate the slope of a straight-line regression for two variables with a nonlinear relationship. It is up to you to make sure that you think through the problem at hand, choose the correct PROC, and carefully specify the SAS variables used in the statistical PROC you choose. The services of a trained statistician may be necessary to help you do this.

The Roadmap

Univariate Statistics

Tabular frequency distributions for variables at any level of measurement may be obtained from PROC FREQ. Bar charts of frequency distributions may be obtained from PROC CHART or PROC GCHART. The frequency distributions of interval and ratio scale variables may be explored in great detail with PROC UNIVARIATE. PROC UNIVARIATE provides an extensive array of descriptive statistics such as mean variance skewness and kurtosis and distribution quantiles including the median. It also provides graphic depictions of variable distributions such as Tukey stem-and-leaf plots, box plots, and normal probability plots. PROC MEANS produces fewer descriptive statistics than PROC UNIVARIATE, but is useful for printing means, standard deviations, and related inferential statistics in a tabular layout.

Simple Bivariate Tests and Measures of Association

PROC CHART or PROC GCHART may be used to produce side-by-side vertical or horizontal bar charts that can help you visualize the relationship between a categorical grouping variable and another variable. PROC PLOT and PROC GPLOT are useful for plotting the relationship between two continuous variables. PROC GPLOT can also add fitted regression lines and confidence bands to plots of data points. It is also capable of fitting curves to data points based on polynomial regression and other, more sophisticated methods such as Lagrange interpolation and cubic splines.

In addition to frequency counts for values of single variables, PROC FREQ can also produce joint frequency distributions, or crosstabulations, of two categorical variables. It can also provide many useful tests and measures of association that assess the existence and strength of relationship between them. In particular, it can calculate Pearson, likelihood ratio, and Fisher's exact chi-square statistics. It can also produce various commonly used bivariate measures of association such as Cramer's V, gamma, Kendall's tau-b, Stuart's tau-c, Lambda statistics, phi, and the Spearman rank correlation coefficient. Beginning with Version 6.10, it provides an array of measures of agreement, including Cohen's kappa statistics.

PROC CORR may be used to produce single bivariate correlation coefficients or matrices of bivariate correlation coefficients and related significance tests. It automatically generates the Pearson r, but both Spearman's rho and Kendall's tau-b may be requested. PROC CORR can also produce partial correlation coefficients.

Comparing Means and Analysis of Variance

Bar charts of variable means for categories of one or two grouping variables may be constructed with PROC CHART or GCHART. Box plots of variable frequency distributions within categories of one or more grouping variables may be obtained from PROC UNIVARIATE.

The SAS System provides several PROCs for testing differences among means when all independent variables or factors are fixed effects. If the analysis compares only two groups and a simple between-groups t test will suffice, PROC TTEST may be employed. When differences among several groups defined by one nominal scale variable are being investigated (i.e., one-way ANOVA), either PROC ANOVA or PROC GLM is appropriate. When two or more factors are involved in the analysis (i.e., 2-way to *n*-way ANOVA), the choice of procedure is generally governed by whether or not the design is balanced. (The simplest type of balanced design occurs when each cell has an equal number of observations.) If the design is balanced, PROC ANOVA may be used. PROC GLM is the most generally applicable and recommended PROC, because it provides correct results for both balanced designs and designs with unequal cell frequencies. Both PROC ANOVA and PROC GLM allow the researcher to perform planned or post hoc tests of differences of cell means.

More complicated analyses, involving random effects and repeated measures, can also be performed with the SAS System. PROC NESTED and PROC MIXED may be used for ANOVAs involving random effects or for mixed-model analyses of variance that involve both random and fixed effects. One-way and multiway repeated-measures ANOVAs, with and without fixed effects, between group factors can be performed in PROC GLM using the REPEATED statement. They can also be performed with PROC MIXED if an assumption more or less complicated than sphericity needs to be made about the distribution of errors.

Some situations arise in which the dependent variable for a one-way ANOVA is rank-ordered or measured on an interval or ratio scale that has a badly nonnormal distribution. In these cases, PROC NPAR1WAY may be used to perform tests of group differences based on ranks. These include the Mann-Whitney or Wilcoxon and the Kruskal-Wallis test of differences of ranks as well as two-sample and *k*-sample median, van der Waerden, and Savage tests.

Estimating Prediction Equations with Regression

PROC REG is best used for bivariate and multiple regression analyses. In addition to furnishing estimates of and significance tests for the multiple R^2 and the coefficients of the estimation equation, PROC REG can also produce standardized coefficients, sometimes referred to as Beta coefficients, that are used by social scientists in path models. In addition, PROC REG supplies an excellent range of diagnostic statistics for assessing potential estimation problems due to multicollinearity, misspecification, or influential observations. PROC REG can also provide several varieties of plots of residuals, useful for checking for violations of standard regression assumptions such as linearity and homoscedasticity. If severe mulicollinearity is found among the independent variables, PROC ORTHOREG can provide more accurate parameter estimates than PROC REG. Use PROC NLIN when the regression equation to be estimated has nonlinear coefficients and nonlinear regression estimates are needed.

The SAS System includes several PROCs that are most appropriate when the regression model to be estimated involves data collected over time. This is often the case for many models involving economic variables. When the errors for the regression model can be assumed to follow an autoregressive process, PROC AUTOREG can be used. If they are assumed to follow a distributed lag model, PROC PDLREG provides appropriate estimates. PROC TSCSREG may be employed when the data arise from a time-series, cross-sectional design, often referred to as a panel study.

Specialized Regression Models

The SAS System provides procedures for estimating the parameters of specialized regression models for dependent variables that are not interval or ratio scale. If the dependent variable has just two values, such as "voted Democrat" versus "voted Republican," or "contracted a certain disease" versus "does not have that disease," and you want coefficient estimates of a logistic regression model and associated odds ratios, use PROC LOGISTIC. PROC LOGISTIC can also be used to estimate a probit regression model for a dichotomous dependent variable as well as probit or logistic regression models for dependent variables with three or more ordered categories. PROC CATMOD may be employed to estimate logistic regression models for dependent variables with three or more unordered levels.

Researchers developing survival models predicting time to component failure, birth of first child, death, or some other event will find three useful PROCs in the SAS System. Those who need to create models using the Cox proportional hazards technique will find it in PROC PHREG. Those who desire to use the parametric accelerated failure time approach to modeling events will prefer PROC LIFEREG. PROC LIFETEST may be used for estimating survival distribution functions and related functions, comparing them for different groups of observations, and identifying variables associated with them.

Poisson regression models for count data and multinomial logistic models for predicting choices among related alternatives may be estimated with PROC PHREG. Tobit regression analysis for both left and right censored variables can be performed with PROC LIFEREG. Loglinear model analysis of crosstabulated categorical variables can be done with PROC CATMOD.

Models for More than One Dependent Variable: Multivariate Statistics

Multivariate descriptive and inferential statistics are usually employed when more than one variable is viewed as dependent and they are intercorrelated. These statistics examine associations between the sets of independent and dependent variables. PROC REG and PROC GLM can handle multivariate regression and ANOVA and perform multivariate tests of significance for single or multiple independent variables. PROC

CANCORR is useful for obtaining canonical correlation analyses of two sets of variables.

Several SAS procedures are useful when a causal structure is assumed among the dependent variables. This is common in econometric or structural equation modeling. Simple recursive systems of equations may be estimated with PROC REG, and its standardized regression coefficients may be used in path models. PROC SYSLIN may be used to estimate multiple equation econometric models using a variety of estimation techniques. These methods include two-stage least-squares, limited-information maximum-likelihood, three-stage least-squares, full-information maximum-likelihood, iterated three-stage least-squares, K-class, and minimum expected loss. PROC SYSLIN can also be used to apply Zellner's seemingly unrelated regressions and iterated seemingly unrelated regressions methods.

Linear structural equation systems with latent variables may be constructed with PROC CALIS. These systems are sometimes referred to as LISREL® models, and are used extensively in education, political science, psychology, and sociology. Although this PROC offers several notations for specifying structural equation models, the LINEQS notation is perhaps the easiest to understand and corresponds very closely to the EQS notation originated by Peter Bentler. PROC MODEL may be employed to create very complex equation systems involving nonlinear parameters for forecasting in economics and other areas.

Measurement and Scaling Techniques

The SAS System offers a variety of PROCs that are useful in measurement and scaling. PROC CORR may be used with the ALPHA option to obtain Cronbach's alpha measure of the reliability of an additive scale as well as item-to-total correlations. PROC PRINCOMP performs principal component analysis using raw data or correlation or covariance matrices. PROC FACTOR provides a very complete implementation of a wide variety of exploratory factor analytical extraction and rotation techniques. Confirmatory factor analysis for a single group of observations can be performed with PROC CALIS. PROC CALIS, however, cannot test for equivalent factor structures across groups of observations.

Two procedures that are useful for obtaining spatial representations of data are PROC CORRESP and PROC MDS. PROC CORRESP performs a correspondence analysis for a crosstabulation of two or more categorical variables. It provides a graphical depiction of the relationship among the variables by representing each row or column of their crosstabulation as a point in a Euclidean space. PROC MDS may also be used to array objects in a Euclidean space by performing two- and three-way, metric and nonmetric multidimensional scaling on matrices of measures of dissimilarities between objects.

Grouping and Predicting Group Membership

Several procedures are available in SAS for grouping observations or variables using cluster analytical methods. PROC CLUSTER uses a hierarchical algorithm to cluster observations. PROC FASTCLUS is useful for disjoint clustering of large numbers of observations. PROC MODECLUS clusters observations using a nonparametric density technique. Variables may be clustered with PROC VARCLUS. PROC TREE is used for printing tree diagrams of cluster formation for PROC CLUSTER or PROC VARCLUS.

Discriminant classification functions for predicting membership in groups from a number of normally distributed, continuous variables can be estimated with PROC DISCRIM. Stepwise development of such functions can be done with PROC STEPDISC. Canonical discriminant function analysis may be carried out with PROC CANDISC.

Navigating the Documentation Maze

In the chapters that follow, we will give you an introduction to many of the most commonly used SAS statistical procedures. However, we will not be able to cover all the potentially useful PROCs mentioned in this chapter. We won't even be able to illustrate every capability of the PROCs we will discuss. To gain a more complete picture of SAS's statistical PROCs or to understand the details of any single PROC, you must consult the documentation for them produced by SAS Institute. The problem for most users is that this documentation is voluminous and is not contained completely in any one publication. In fact, given the large number of different manuals produced by SAS Institute, finding the documentation for a particular PROC can be like navigating a maze. Exhibit 7.1, below, will help you navigate this documentation maze.

The first thing you must understand to use Exhibit 7.1 is that the SAS System is not a single software product. Rather, it is composed of many groups of software modules or "products." The statistical PROCs referred to in this chapter and later in this book are each contained in one of these products: Base SAS, SAS/STAT, SAS/ETS, and SAS/GRAPH. (They form the main column headings of the table.) The basic documentation for the statistical PROCs mentioned in this book for each product is listed in Exhibit 7.2. In the future, as SAS Institute expands and revises its software, the SAS version or edition numbers for some of the guides and references may change. However, the basic titles should remain the same. The documents referred to below should provide you with information that is current through Release 6.10 for the PROCs listed in the table.

In the body of the table in Exhibit 7.1, we have listed the names of SAS statistical procedures used for the types of statistical analysis listed in the row headings. For each PROC, we have given the chapter number in the corresponding SAS product documentation that discusses that PROC. When you need more information about a statistical PROC, this table should provide a handy guide to tracking down the basic SAS Institute reference for it.

Exhibit 7.1

Documentation for SAS PROCs for Statistical Analysis

Statistical Technique	Base SAS PROC/Chapter	SAS/STAT PROC/Chapter	SAS/ETS PROC/ Chapter	SAS/GRAPH PROC/ Chapter
Univariate Statistics	CHART/9 FREQ/20 MEANS/21 UNIVARIATE/42	FREQ/23		GCHART/23
Bivariate Statistics	CORR/15 FREQ/20 PLOT/25	NPAR1WAY/30 TTEST/42		GCHART/23 GPLOT/31
ANOVA/ANCOVA		ANOVA/13 GLM/24 MIXED/16, P229 NESTED/28		
Linear Regression		ORTHOREG/31 REG/36	AUTOREG/4 PDLREG/13 TSCSREG/18	
Nonlinear Regression		NLIN/29		
Categorical DV Regression		LOGISTIC/27 CATMOD/17		
Truncated DV Regression (Tobit)		LIFEREG/25		
Survival Analysis		LIFEREG/25 LIFETEST/26 PHREG/19, P229		
Multivariate		CANCORR/15 PRINCOMP/33		
Structural Equations/ Econometric Modeling		CALIS/14	MODEL/11 SYSLIN/17	
Reliability/ Measurement/ Scaling	CORR/15	CALIS/14 CORRESP/19 FACTOR/21 MDS/15, P229		
Grouping/ Clustering/ Discriminant Analysis		FASTCLUS/22 MODECLUS/P256 STEPDISC/39 TREE/41 VARCLUS/43		

Documentation

Exhibit 7.2 identifies SAS Institute documents that are relevant to the procedures described in this chapter. Chapter numbers are for the first publication for each product cited below. Chapter numbers followed by a P229 or P256 refer to chapters in the corresponding *SAS Technical Reports*, below. See Appendix B for ordering information and a description of other publications available from SAS Institute.

Exhibit 7.2

SAS Institute Statistical Documentation

Base SAS:	*SAS® Procedures Guide, Version 6, Third Edition*
SAS/STAT:	*SAS/STAT® User's Guide, Version 6, Fourth Edition*
	SAS® Technical Report P-229, SAS/STAT Software: Changes and Enhancements, Release 6.07
	SAS® Technical Report P-256, SAS/STAT Software: The MODECLUS Procedure, Release 6.09
	SAS/STAT® Software: Changes and Enhancements, Release 6.10
SAS/ETS:	*SAS/ETS® User's Guide, Version 6, Second Edition*
SAS/GRAPH:	*SAS/GRAPH® Software Reference, Version 6, First Edition, Volumes 1 and 2*
	SAS/GRAPH® Software: Introduction, Version 6, First Edition

Recap

As you can see, SAS can perform almost any type of statistical analysis you may need, ranging from the simple to the highly complex. We have covered a very broad array of statistical techniques implemented in SAS PROCs in this chapter so that it would be useful for identifying the PROCs required for basic as well as advanced research. In the chapters that follow, we will examine in more depth some of the most generally used PROCs mentioned above.

8

Statistics for Single Variables

In this chapter, you will learn how to produce many of the tabular displays, graphs, and descriptive statistics commonly used to summarize the values of a single variable. We will deal with two general types of variables — categorical and continuous. A *categorical* variable has a relatively small number of discrete values. You may use PROC FREQ to obtain a frequency distribution of the values of such a variable. You may produce bar graphs of the frequency distribution of a categorical variable by using PROC CHART or PROC GCHART.

If the variable of interest is *continuous* and has many numeric values, use PROC MEANS to get the variable's summary statistics, such as the mean and the standard deviation. If you need more complete information on the distribution of a continuous variable, such as the median, skewness, or interquartile range, use PROC UNIVARIATE. PROC CHART can be used to produce histograms for the distributions of continuous variables.

Examining Distributions of Categorical Variables

Creating Frequency Distributions

One way to summarize the values of a categorical variable is to count the number of occurrences of each value of the variable. A tabular presentation of a variable's values and their frequencies is usually called a frequency distribution. PROC FREQ may be used to generate frequency distributions for one or more categorical variables in a SAS dataset. The code in Exhibit 8.1 uses a small random sample of data on individuals, taken from the 1991 General Social Survey (GSS) conducted by the

National Opinion Research Center at the University of Chicago. The program creates SAS dataset GSS1991, used in for most of the examples in this chapter.

Exhibit 8.1

Read GSS Data and Save SAS Dataset.

Program

```
DATA  gss1991 ;

/* Use list input. Be careful to add "&" for    */
/* region variable because of embedded blanks. */
INPUT   idnum   cregion & $    age    marital ;

/* Label SAS variable names. */
LABEL   idnum     = 'Respondent Identification Number'
        cregion   = 'Character U.S. Census Region Name'
        age       = 'Age of Respondent'
        marital   = 'Respondent Marital Status'  ;
                /*  1 =  Married     2 = Divorced
                    3 = Widowed      4 = Separated
                    5 = Single */
DATALINES;
  1        Mid. Atlantic       33        1
  2        Pacific             60        1
  3        Pacific             37        5
  4        E. N. Central       58        2
  5        E. N. Central       50        1
  6        Mid. Atlantic       36        1
  7        Pacific             31        4
  8        Sou. Atlantic       29        3
  9        W. S. Central       23        5
 10        New England         40        1
 11        W. N. Central       71        1
 12        Sou. Atlantic       57        3
 13        Sou. Atlantic       36        1
 14        Mid. Atlantic       36        1
 15        New England         30        5
 16        New England         44        1
 17        Mid. Atlantic       65        1
 18        E. N. Central       19        5
 19        E. N. Central       33        5
 20        W. N. Central       47        1
 21        E. N. Central       47        1
 22        E. N. Central       74        1
 23        E. N. Central       44        1
 24        E. S. Central       44        1
 25        Sou. Atlantic       32        3
 26        Sou. Atlantic       67        2
 27        Sou. Atlantic       29        1
 28        W. S. Central       42        1
 29        E. S. Central       78        2
 30        W. S. Central       47        2
 31        E. S. Central       66        2
 32        Sou. Atlantic       35        5
 33        Pacific             77        2
 34        Mountain            23        5
 35        Pacific             26        5
 36        Pacific             21        5
```

Exhibit 8.1

Read GSS Data and Save SAS Dataset. (continued)

```
 37      Pacific          71     1
 38      Mountain         20     5
 39      Mid. Atlantic    74     2
 40      New England      58     2
 41      E. N. Central    39     5
 42      W. N. Central    20     5
 43      E. N. Central    66     2
 44      W. N. Central    37     1
 45      E. S. Central    36     1
 46      W. S. Central    58     1
 47      Sou. Atlantic    37     4
 48      W. S. Central    34     1
 49      E. S. Central    55     2
 50      Sou. Atlantic    22     5
 51      Sou. Atlantic    33     1
 52      Pacific          66     2
RUN;
```

Basic PROC FREQ Syntax Exhibit 8.2 illustrates how PROC FREQ displays a frequency distribution when it is used without any options. The TABLES statement is placed immediately after the PROC FREQ statement. It lists the names of variables for which you want frequency distributions. A separate frequency distribution will be created for each variable listed on the TABLES statement. The variables may be character or numeric. Be careful not to list continuous variables with many values — this can result in long, useless lists of values with frequencies of 1. Always use a TABLES statement! Without one, PROC FREQ will produce a frequency distribution for *every* variable in the dataset.

Look at the output in Exhibit 8.2. The variable label supplied in the DATA step (Exhibit 8.1) is printed above each variable's frequency distribution. The leftmost column of the frequency distribution table for each variable is labeled with the name of the variable being summarized. It contains a list of the variable's values. The values of long character variables, such as `cregion`, are truncated to eight characters. The next column, labeled `Frequency`, gives the number of observations for a given value of the variable. The `Percent` column gives the percentage of the total for a given value. The percentage is based on the number of nonmissing observations in the dataset.

It is important to remember that PROC FREQ does not use observations with missing values unless it is explicitly told to do so. The MISSING option in the TABLES statement is described later.

The `Cumulative Frequency` column gives a running total of the counts; the `Cumulative Percent` column gives each cumulative frequency's percentage of the total number of cases. Note that the last cumulative frequency is the total number of cases without missing values. Since these data contained no missing values, 52 is the total number of observations in the dataset.

Exhibit 8.2

Basic Syntax and Output for PROC FREQ

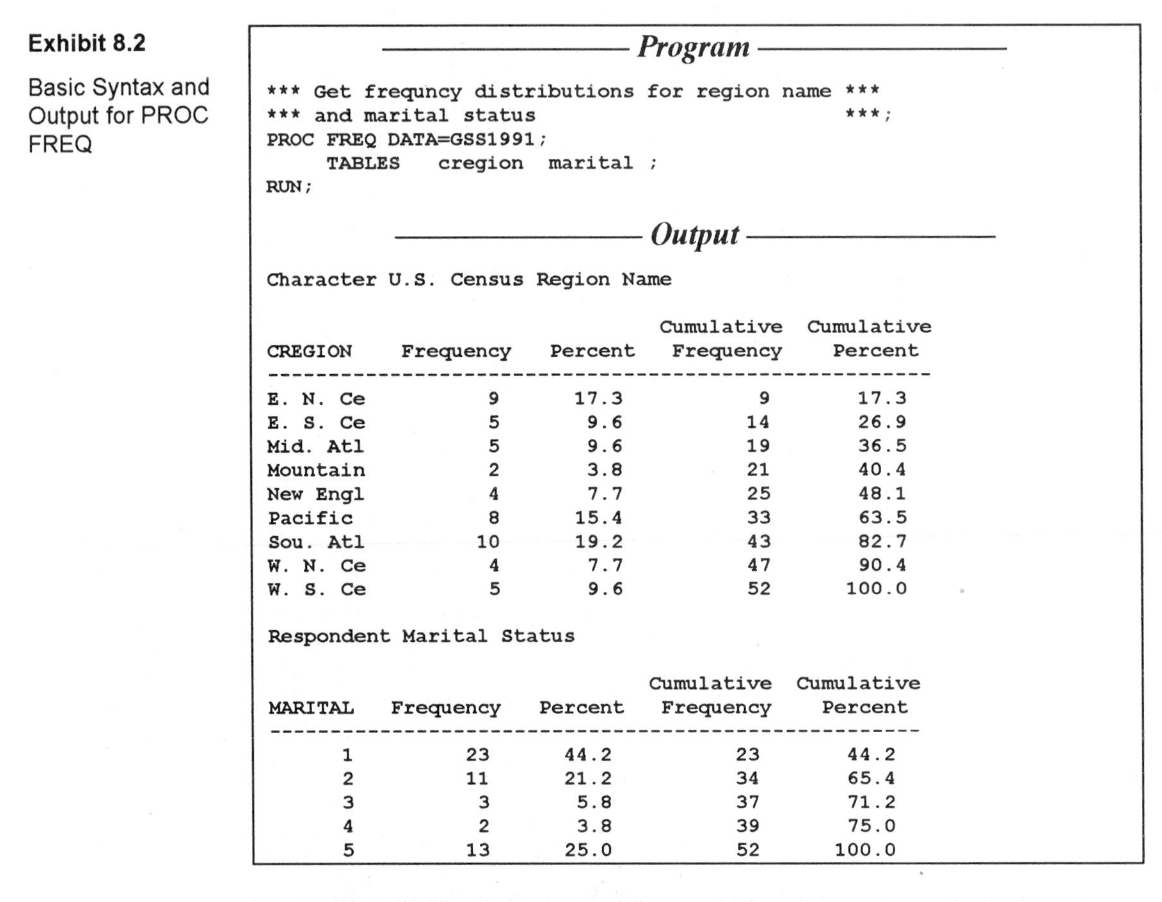

Program

```
*** Get frequncy distributions for region name ***
*** and marital status                          ***;
PROC FREQ DATA=GSS1991;
     TABLES   cregion   marital ;
RUN;
```

Output

```
Character U.S. Census Region Name

                                         Cumulative  Cumulative
CREGION     Frequency    Percent     Frequency     Percent
-------------------------------------------------------------
E. N. Ce            9       17.3             9        17.3
E. S. Ce            5        9.6            14        26.9
Mid. Atl            5        9.6            19        36.5
Mountain            2        3.8            21        40.4
New Engl            4        7.7            25        48.1
Pacific             8       15.4            33        63.5
Sou. Atl           10       19.2            43        82.7
W. N. Ce            4        7.7            47        90.4
W. S. Ce            5        9.6            52       100.0

Respondent Marital Status

                                        Cumulative  Cumulative
MARITAL    Frequency    Percent     Frequency     Percent
-------------------------------------------------------------
      1          23       44.2            23        44.2
      2          11       21.2            34        65.4
      3           3        5.8            37        71.2
      4           2        3.8            39        75.0
      5          13       25.0            52       100.0
```

Controlling Table Content and Presentation You may use the ORDER= option on the PROC FREQ statement to control the order in which the values of a variable are printed in the frequency distribution table. If you do not use this option, a variable's numeric values are printed from lowest to highest and character variables are printed alphabetically. To arrange the values of a variable in the order of their frequency, from most frequent to least frequent, use ORDER=FREQ. This option can be very useful when analyzing survey data or other data generated when respondents answer questions involving choices among alternatives.

If you would like to handle missing values as though they were not missing, use the MISSING option of the TABLES statement. This will include observations with missing values in the table. The NOCUM option may be used to suppress the printing of the cumulative frequencies and percentages. Either or both of these options are placed after a slash (/) following the variable list on the TABLES statement.

The program in Exhibit 8.3 uses the ORDER option to arrange the values of the census region variable by number of respondents in each region. The NOCUM option suppresses the printing of the cumulative frequencies and cumulative percentages.

The program also adds a title to the output. This example assumes that the SAS dataset `gss1991`, created in the preceding example, has been saved as a permanent SAS dataset in a previous SAS program.

Look at the Output portion of Exhibit 8.3. Notice that the South Atlantic region is printed first in this frequency table because it had the highest number of respondents.

Exhibit 8.3

PROC FREQ Using the ORDER= and NOCUM Options

Program

```
LIBNAME  data  'location';

PROC FREQ  DATA=data.gss1991  ORDER=FREQ ;
     TABLES   cregion  /  NOCUM ;
TITLE 'Census Regions Ordered by Number of Respondents' ;
RUN;
```

Output

```
Census Regions Ordered by Number of Respondents

Character U.S. Census Region Name

CREGION    Frequency   Percent
------------------------------
Sou. Atl          10      19.2
E. N. Ce           9      17.3
Pacific            8      15.4
E. S. Ce           5       9.6
Mid. Atl           5       9.6
W. S. Ce           5       9.6
New Engl           4       7.7
W. N. Ce           4       7.7
Mountain           2       3.8
```

Visualizing Frequency Distributions with Bar Charts

It is often easier to visualize a distribution using a bar chart than it is to examine counts of values in a frequency distribution. Bar charts may be constructed with horizontal or vertical bars. In either form, the length of the bar for each value of the analysis variable corresponds to the raw frequency or percentage of cases having that value in the dataset. PROC CHART uses printed characters to construct the bars in either in the Display Manager OUTPUT window or on an impact, inkjet, or laser printer.

PROC GCHART, which is available in SAS Institute's SAS/GRAPH product, may be used to produce higher-resolution bar charts on a computer screen, a pen plotter, an inkjet or laser printer, or even a slide. The basic syntax necessary to produce bar charts for PROC CHART and PROC GCHART is quite similar. Since almost everyone has access to a character printer, the discussion below will focus on PROC CHART. GCHART may be substituted for CHART in most cases, especially if you are using SAS in interactive mode and wish the graphic output to be displayed on your monitor.

Basic CHART Syntax The program in Exhibit 8.4 shows how simple it is to produce bar charts. It uses the minimal amount of syntax to request both a vertical and a horizontal bar chart for the variable `cregion` in our `gss1991` dataset. At least one HBAR or VBAR statement is required. Multiple chart requests are allowed. You may request charts for more than one variable by including multiple variable names after HBAR or VBAR.

Examine the Output portion of Exhibit 8.4. Note that the horizontal bar chart produced by the HBAR statement automatically creates a frequency distribution table for `state` to the right of the horizontal bars. Note also that PROC CHART truncates long character values to eight characters and, for vertical bar charts, prints their values vertically.

Exhibit 8.4

Default Vertical and Horizontal Bar Chart from PROC CHART

Program

```
LIBNAME  data  'location' ;

PROC CHART DATA=data.gss1991 ;
     HBAR   cregion ;
     VBAR   cregion ;
run;
```

Output

```
CREGION                                          Cum.                Cum.
                                          Freq   Freq   Percent   Percent
              |
E. N. Ce      |******************           9      9     17.31     17.31
              |
E. S. Ce      |**********                   5     14      9.62     26.92
              |
Mid. Atl      |**********                   5     19      9.62     36.54
              |
Mountain      |****                         2     21      3.85     40.38
              |
New Engl      |********                     4     25      7.69     48.08
              |
Pacific       |****************             8     33     15.38     63.46
              |
Sou. Atl      |********************        10     43     19.23     82.69
              |
W. N. Ce      |********                     4     47      7.69     90.38
              |
W. S. Ce      |**********                   5     52      9.62    100.00
              |
              ----+---+---+---+---+
                  2   4   6   8  10
                   Frequency
```

Exhibit 8.4

Default Vertical and Horizontal Bar Chart from PROC CHART (continued)

New Page of Output

```
Frequency

10 +                                                                    *****
   |  *****                                                             *****
   |  *****                                                  *****      *****
   |  *****                                                  *****      *****
   |  *****                                                  *****      *****
 5 +  *****      *****      *****                            *****      *****                 *****
   |  *****      *****      *****                 *****      *****      *****      *****      *****
   |  *****      *****      *****                 *****      *****      *****      *****      *****
   |  *****      *****      *****      *****      *****      *****      *****      *****      *****
   |  *****      *****      *****      *****      *****      *****      *****      *****      *****
   ------------------------------------------------------------------------------------------------------
        E          E          M          M          N          P          S          W          W
        .          .          i          o          e          a          o          .          .
                              d          u          w          c          u
        N          S          .          n                     i          .          N          S
        .          .                     t          E          f                     .          .
                              A          a          n          i          A
        C          C          t          i          g          c          t          C          C
        e          e          l          n          l                     l          e          e

                                 Character U.S. Census Region Name
```

Charting Discrete Numeric Variables PROC CHART has a number of options that can be applied to both HBAR and VBAR statements. These indicate how you wish to treat numeric and missing values, and how you want to arrange the bars from left to right. One or more of these options can be placed after a "/" following the list of variable names on the HBAR or VBAR statement.

When the variable list contains a numeric, variable that you wish to treat as categorical, you must use the DISCRETE option. It tells PROC CHART that you want numeric values to be treated as discrete categories rather than as values of a continuous variable. If you do not use this option, PROC CHART will assume that numeric values are from continuous variables. It will group values of the variable into bars whose labels may not correspond exactly to the numeric values you wish to see.

Look at the program in Exhibit 8.5 to see how employing the DISCRETE option produces very different bar labels for the numeric variable for the variable `marital`, which, as you have seen in Exhibit 8.2, has only five discrete values.

In Exhibit 8.5's Output section, compare the values for the bar labels for the first vertical bar chart with those for the second. Clearly, using the DISCRETE option makes the second bar chart more appropriate for this variable.

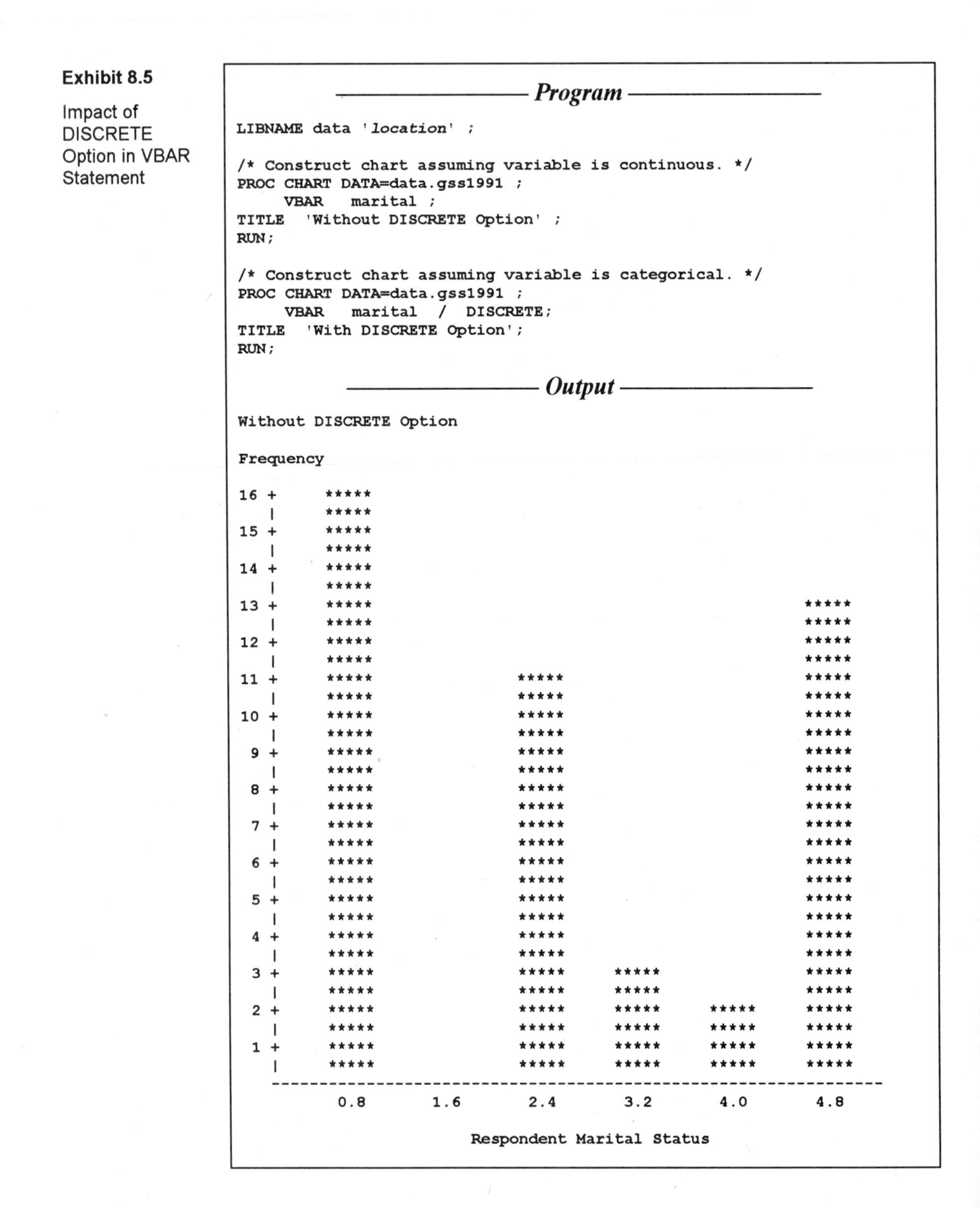

Exhibit 8.5

Impact of DISCRETE Option in VBAR Statement

Program

```
LIBNAME data 'location' ;

/* Construct chart assuming variable is continuous. */
PROC CHART DATA=data.gss1991 ;
     VBAR   marital ;
TITLE  'Without DISCRETE Option' ;
RUN;

/* Construct chart assuming variable is categorical. */
PROC CHART DATA=data.gss1991 ;
     VBAR   marital   /  DISCRETE;
TITLE  'With DISCRETE Option';
RUN;
```

Output

```
Without DISCRETE Option

Frequency

16 +     *****
   |     *****
15 +     *****
   |     *****
14 +     *****
   |     *****
13 +     *****                                             *****
   |     *****                                             *****
12 +     *****                                             *****
   |     *****                                             *****
11 +     *****               *****                         *****
   |     *****               *****                         *****
10 +     *****               *****                         *****
   |     *****               *****                         *****
 9 +     *****               *****                         *****
   |     *****               *****                         *****
 8 +     *****               *****                         *****
   |     *****               *****                         *****
 7 +     *****               *****                         *****
   |     *****               *****                         *****
 6 +     *****               *****                         *****
   |     *****               *****                         *****
 5 +     *****               *****                         *****
   |     *****               *****                         *****
 4 +     *****               *****                         *****
   |     *****               *****                         *****
 3 +     *****               *****     *****               *****
   |     *****               *****     *****               *****
 2 +     *****               *****     *****     *****     *****
   |     *****               *****     *****     *****     *****
 1 +     *****               *****     *****     *****     *****
   |     *****               *****     *****     *****     *****
   ----------------------------------------------------------------
          0.8       1.6       2.4       3.2       4.0       4.8

                        Respondent Marital Status
```

Exhibit 8.5

Impact of DISCRETE Option in VBAR Statement (continued)

New Page of Output

```
With DISCRETE Option

Frequency

23 +     *****
   |     *****
22 +     *****
   |     *****
21 +     *****
   |     *****
20 +     *****
   |     *****
19 +     *****
   |     *****
18 +     *****
   |     *****
17 +     *****
   |     *****
16 +     *****
   |     *****
15 +     *****
   |     *****
14 +     *****
   |     *****
13 +     *****                                         *****
   |     *****                                         *****
12 +     *****                                         *****
   |     *****                                         *****
11 +     *****       *****                             *****
   |     *****       *****                             *****
10 +     *****       *****                             *****
   |     *****       *****                             *****
 9 +     *****       *****                             *****
   |     *****       *****                             *****
 8 +     *****       *****                             *****
   |     *****       *****                             *****
 7 +     *****       *****                             *****
   |     *****       *****                             *****
 6 +     *****       *****                             *****
   |     *****       *****                             *****
 5 +     *****       *****                             *****
   |     *****       *****                             *****
 4 +     *****       *****                             *****
   |     *****       *****                             *****
 3 +     *****       *****       *****                 *****
   |     *****       *****       *****                 *****
 2 +     *****       *****       *****       *****     *****
   |     *****       *****       *****       *****     *****
 1 +     *****       *****       *****       *****     *****
   |     *****       *****       *****       *****     *****
   -----------------------------------------------------------------
           1           2           3           4           5

                    Respondent Marital Status
```

Controlling Bar Content and Order Normally, the bar charts produced by PROC CHART do not include a bar for SAS's missing value. If you would like to include a bar representing missing values, use the MISSING option on either the HBAR or the VBAR statement. If you use the MISSING option, any calculations of percentages and cumulative totals produced will include observations with missing values. Its effect is similar to that of the MISSING option in PROC FREQ's TABLES statement, described earlier.

You may wish to change the default ordering of the bars from top to bottom or left to right. By default, PROC CHART arranges the bars for numeric variables based on the numeric values they represent. It places the values for character variables in alphabetical sorting order. Some researchers find it useful to arrange bars according to the size of the frequency counts they represent. Use either the ASCENDING or the DESCENDING option to arrange the bars in ascending or descending order of their frequency counts. These options can be very useful when you are dealing with variables representing choices made by people and the bar chart needs to reflect the relative popularity of a set of alternative choices.

Controlling Output Appearance The SPACE= option may be used to control spacing between bars on the HBAR and VBAR statements. Follow the equal sign with a number equal to the number of blank spaces you would like between bars. Using SPACE=0 will eliminate any blank spaces between bars and may be used to produce histograms.

The options we have discussed so far may be added to either an HBAR or a VBAR statement. NOSTATS is an option unique to the HBAR statement. Adding it to an HBAR statement will suppress the frequency distribution printed to the right of the bar chart.

Let's see how some of these options work when we use them to construct some bar charts for the marital status variable. Look at the program in Exhibit 8.6. The first PROC CHART uses the DESCENDING and NOSTATS options when constructing a horizontal bar graph. The second employs the ASCENDING and NOSPACE options for constructing a vertical bar graph.

Exhibit 8.6

Use DISCRETE, DESCENDING, ASCENDING, NOSTATS, and SPACE= Options.

Program

```
LIBNAME  data  'location' ;
PROC CHART DATA=data.gss1991 ;
     HBAR  marital  /  DISCRETE DESCENDING NOSTATS ;
TITLE  'Use DISCRETE, DESCENDING, and NOSTATS Options';
RUN;

PROC CHART DATA=data.gss1991 ;
     VBAR  marital  /  DISCRETE ASCENDING SPACE=0 ;
TITLE1  'VBAR with ASCENDING and SPACE=0 Options' ;
RUN;
```

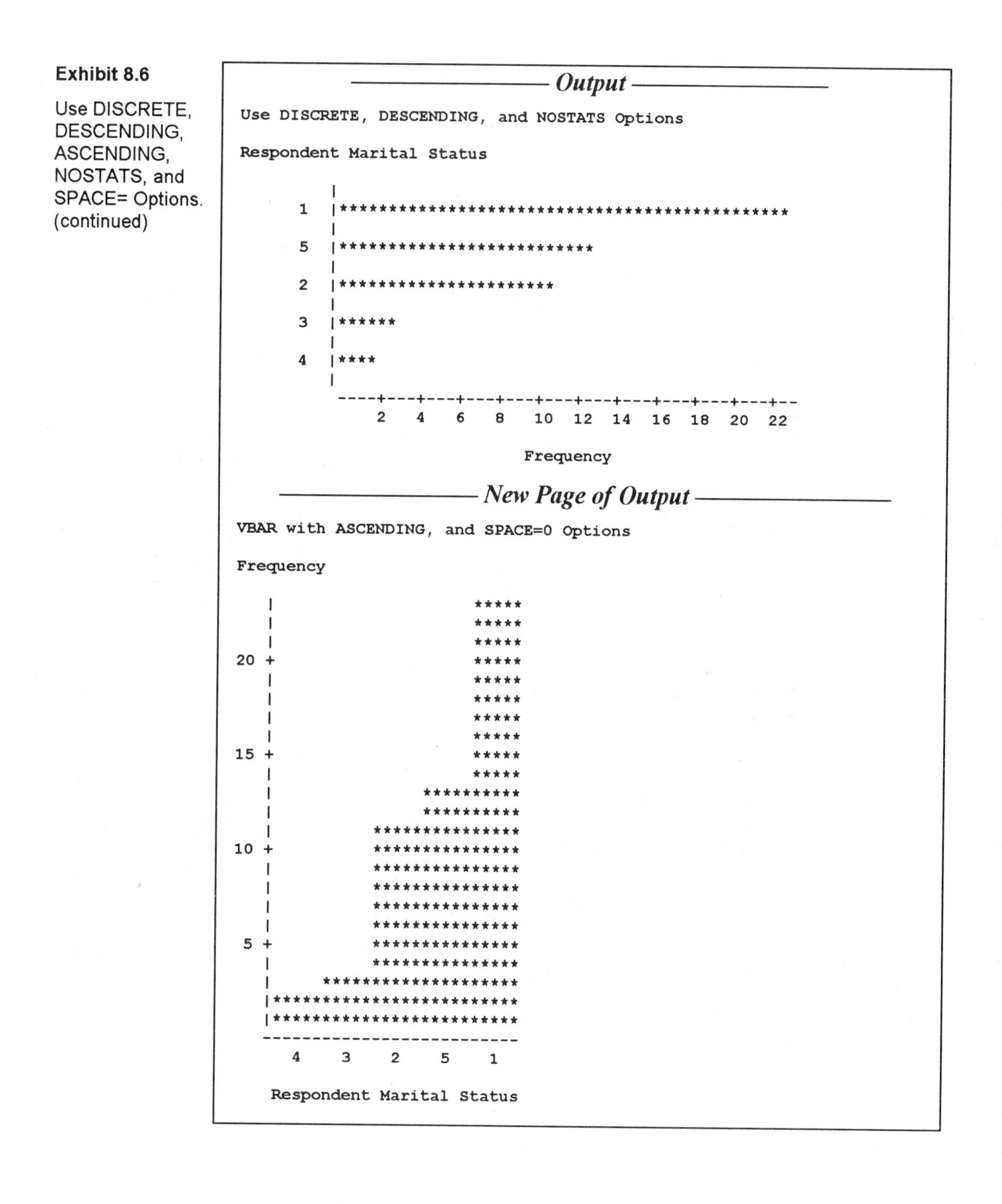

Exhibit 8.6

Use DISCRETE, DESCENDING, ASCENDING, NOSTATS, and SPACE= Options. (continued)

Output

```
Use DISCRETE, DESCENDING, and NOSTATS Options

Respondent Marital Status

         |
    1    |**********************************************
         |
    5    |**************************
         |
    2    |**********************
         |
    3    |******
         |
    4    |****
         |
          ----+---+---+---+---+---+---+---+---+---+---+--
              2   4   6   8  10  12  14  16  18  20  22

                            Frequency
```

New Page of Output

```
VBAR with ASCENDING, and SPACE=0 Options

Frequency

    |                    *****
    |                    *****
    |                    *****
 20 +                    *****
    |                    *****
    |                    *****
    |                    *****
    |                    *****
 15 +                    *****
    |                    *****
    |               **********
    |               **********
    |          ***************
 10 +          ***************
    |          ***************
    |          ***************
    |          ***************
    |          ***************
  5 +          ***************
    |          ***************
    |     ********************
    |*************************
    |*************************
    --------------------------
       4    3    2    5    1

       Respondent Marital Status
```

Because continuous variables can have many unique values, summarizing their distributions with tools that enumerate every value is usually uninformative and wasteful of computing and paper resources. For this reason, frequency distributions for continuous variables generally group observations into ranges of values and then give frequency counts of the number of cases within each range. Bar charts called *histograms* are constructed using this strategy. PROC CHART and PROC GCHART are useful for graphically displaying histograms for continuous variables.

In addition to visual displays of frequency distributions, a variety of descriptive statistics may be calculated that help to summarize the central tendency and variability of continuous variables. PROC MEANS may be used to obtain basic descriptive statistics such as the mean and the standard deviation. More detailed descriptive statistics and information about the frequency distributions of continuous variables can be obtained from PROC UNIVARIATE.

Examining Distributions of Continuous Variables

Remember that PROC CHART assumes that any numeric SAS variables listed on an HBAR or VBAR statement are continuous unless the DISCRETE option is present. When it is absent, CHART constructs a histogram for each of the variables listed. You can use SAS code similar to that presented in Exhibit 8.7 to obtain a histogram for a continuous variable. In this case, we're looking at the distribution of the variable `age` from the `gss1991` data listed at the beginning of this chapter.

If you do not specify either the number of bars or their midpoints, CHART uses an algorithm to select the number of bars, their value ranges, and their midpoints when it constructs a histogram for a continuous variable. In the chart in Exhibit 8.7, it used value ranges of 10 years. The numerical labels to the left of the bars are the *midpoints* of the value ranges — not their beginning or ending values. Thus, the bar labeled 20 includes the seven cases with ages less than 24, the bar labeled 30 includes all observations with an age value greater than or equal to 25 and less than 35, and so on.

Specifying the Chart's Numeric Intervals Sometimes you want to exercise control over either the number of bars or their value ranges and midpoints. To specify the midpoints of value ranges and let PROC CHART work out their upper and lower bounds, use a "/", followed by the MIDPOINTS= option, in a VBAR or HBAR statement. Simply follow the equal sign with a list of the values of the midpoints, separated by at least one space, as in MIDPOINTS= 15 25 35 45 55 65 75. Alternatively, you may generate a series of midpoints by using syntax such as MIDPOINTS = 15 TO 75 BY 10. This would generate bars with the same midpoints as those in the preceding list of values.

Exhibit 8.7

Using PROC CHART to Produce a Histogram

Program

```
LIBNAME data 'location' ;

PROC CHART DATA=data.gss1991 ;
     HBAR  age ;
TITLE  'Producing a Histogram Using No Options';
RUN;
```

Output

```
Producing a Histogram Using No Options

Age of Respondent                                      Cum.              Cum.
Midpoint                                         Freq  Freq  Percent  Percent
         |
 20      |**************                            7     7    13.46    13.46
         |
 30      |********************                     10    17    19.23    32.69
         |
 40      |****************************             14    31    26.92    59.62
         |
 50      |********                                  4    35     7.69    67.31
         |
 60      |************                              6    41    11.54    78.85
         |
 70      |******************                        9    50    17.31    96.15
         |
 80      |****                                      2    52     3.85   100.00
         |
         ----+---+---+---+---+---+---+
             2   4   6   8  10  12  14

                   Frequency
```

It is generally statistically advisable to use, and easier to interpret, equally spaced midpoints when dealing with continuous numeric variables, but there are situations in which this is not the case. For example, frequencies for some data might better be displayed on a logarithmic frequency scale using MIDPOINTS= 10 100 1000 10000 100000. PROC CHART does not require that the midpoint values be equally spaced.

The program in Exhibit 8.8 illustrates the use of the MIDPOINTS= option by producing a bar chart for `age` with different midpoints. You can see the result of specifying the midpoints of the value ranges in this manner by comparing Exhibit 8.7 to Exhibit 8.8.

Exhibit 8.8

Grouping a Continuous Variable

Program

```
LIBNAME data 'location' ;

PROC CHART DATA=data.gss1991 ;
     HBAR  age  /   MIDPOINTS = 15 to 75 by 10 ;
TITLE  'Age of Respondents in 10-Year Intervals';
RUN;
```

Exhibit 8.8

Grouping a Continuous Variable (continued)

Output

```
Age of Respondents in 10-Year Intervals

Age of Respondent                                        Cum.              Cum.
Midpoint                                           Freq  Freq  Percent  Percent
      |
 15   |**                                            1     1     1.92     1.92
      |
 25   |******************                            9    10    17.31    19.23
      |
 35   |********************************             16    26    30.77    50.00
      |
 45   |****************                              8    34    15.38    65.38
      |
 55   |************                                  6    40    11.54    76.92
      |
 65   |************                                  6    46    11.54    88.46
      |
 75   |************                                  6    52    11.54   100.00
      |
      ----+---+---+---+---+---+---+---+
          2   4   6   8  10  12  14  16

                   Frequency
```

If you wish to specify the number of bars and let PROC CHART determine the value ranges and midpoints, use the LEVELS= option. Specify the number of bars after the equal sign. You should exercise caution when choosing the number of bars, because PROC CHART may not select equal-width value ranges or value ranges that make sense given your knowlege of the data. Exhibit 8.9's program illustrates the use of this option — we produce a chart for `age` with five bars.

Exhibit 8.9

Specify Number of Bars Using the LEVELS Option.

Program

```
LIBNAME  data   'location' ;

PROC CHART DATA=data.gss1991 ;
     VBAR  age  /  LEVELS= 5 ;
TITLE1  'Histogram for LEVELS=5';
RUN
```

Exhibit 8.9

Specify Number of Bars Using the LEVELS Option. (continued)

Output

```
Histogram for LEVELS=5
 Frequency
20 +                   *****
   |                   *****
19 +                   *****
   |                   *****
18 +                   *****
   |                   *****
17 +                   *****
   |                   *****
16 +                   *****
   |                   *****
15 +                   *****
   |                   *****
14 +                   *****
   |                   *****
13 +                   *****
   |                   *****
12 +                   *****
   |                   *****
11 +                   *****                   *****
   |                   *****                   *****
10 +                   *****       *****       *****
   |                   *****       *****       *****
 9 +                   *****       *****       *****
   |                   *****       *****       *****
 8 +                   *****       *****       *****
   |                   *****       *****       *****
 7 +                   *****       *****       *****
   |                   *****       *****       *****
 6 +                   *****       *****       *****       *****
   |                   *****       *****       *****       *****
 5 +       *****       *****       *****       *****       *****
   |       *****       *****       *****       *****       *****
 4 +       *****       *****       *****       *****       *****
   |       *****       *****       *****       *****       *****
 3 +       *****       *****       *****       *****       *****
   |       *****       *****       *****       *****       *****
 2 +       *****       *****       *****       *****       *****
   |       *****       *****       *****       *****       *****
 1 +       *****       *****       *****       *****       *****
   |       *****       *****       *****       *****       *****
   ----------------------------------------------------------------
            15          30          45          60          75
                             Age of Respondent
```

PROC MEANS: Basic Summary Statistics for Continuous Variables

PROC MEANS produces a variety of commonly used summary statistics for continuous variables. These include the mean or average, the variance, and the standard deviation. PROC MEANS can also produce the standard error of the mean, useful when data are assumed to be a simple random sample from a larger population. This statistic is used to test the hypothesis that the population mean equals some value. It is also used to set a confidence interval around the sample mean that is highly likely to include the true mean of the population. PROC MEANS also reports the Student's t and its associated level of significance. It tests the hypothesis that the population mean is zero.

Basic MEANS Syntax

Use of PROC MEANS usually involves a PROC statement and a VAR statement. To override the default set of statistics, specify the statistics of interest in the PROC statement. A list of the most commonly used summary statistics and their option names for PROC MEANS is included in Exhibit 8.10. You may place these options in any order on the PROC MEANS statement.

Exhibit 8.10

Summary Statistic Keywords in PROC MEANS

Option Name	Option Description
N	Number of cases used for computing statistics
NMISS	Number of cases with SAS missing values
MEAN	Arithmetic mean, or "average"
VAR	Sample variance (denominator is N - 1)
STD	Sample standard deviation (square root of variance)
STDERR	Standard error of the mean
T	Student's t for testing population mean equal to zero
PRT	Probability of obtaining a larger absolute value for t
CLM	Upper and lower confidence limits for the mean. Default is 95% confidence limits.

Display and Analytical Options

There are three additonal PROC MEANS options. The MAXDEC= option can be used to control the sometimes unwieldy number of decimal places PROC MEANS uses to print its statistics. When this option is used, statistics are rounded to the number of decimal places specified after the equal sign. Keep in mind that most researchers do not use variables whose summary statistics require printing six or seven digits after the decimal point. Using MAXDEC=2 for variables with no decimal values, or with only two or three significant digits, results in more readable output.

The ALPHA= option allows you to specify the alpha probability level for the CLM option in the event that you need limits other than the customary 95% confidence limits. When this option is used, the confidence limits given by PROC MEANS are the 1 - alpha limits. For example, if you use ALPHA=.01, PROC MEANS will report the 99% confidence limits.

Occasionally, you may need the population variance and standard deviation for a variable rather than the corresponding sample statistics. The population statistics are calculated using the actual number of nonmissing valued observations in the variance formula denominator rather than that number less 1 (i.e., N versus N - 1). If the number of observations is over 30, there is little practical difference between the sample and population values of these statistics. However, for a small number of observations the values can be different. If you require population variances and standard deviations, use the VARDEF=N option.

If no statistics are specifically requested on the PROC MEANS statement, the procedure prints the following for each variable: total number of observations (labeled `N Obs`), the number of observations with nonmissing values (labeled `N`), the maximum and minimum values, the mean, and the standard deviation. These statistics, however, are not automatically output when you specifically request other summary statistics on the PROC MEANS statement. Therefore, when you specify the statistics that you want, be sure to provide a complete list.

Specifying Analysis Variables

The names of the variables for which you'd like statistics are listed on a VAR statement following the PROC MEANS statement. Only numeric variables can be included. It is always advisable to use a VAR statement. If the VAR statement is omitted, PROC MEANS will calculate statistics for every numeric variable in the input data set.

Though you'll probably want to tailor the options you use, the example that follows lists summary and inferential statistics that are appropriate for many purposes. In the program in Exhibit 8.11, we included the variable `marital` on the VAR statement to illustrate the fact that MEANS pays no attention to level-of-measurement issues when processing numeric variables. Even though marital status is clearly a nominal scale variable, MEANS calculates all the requested statistics as though it were a continuous variable.

Exhibit 8.11

Specifying Options for PROC MEANS

Program

```
LIBNAME data 'location';
PROC MEANS DATA=data.gss1991    N MEAN STD MIN MAX
     CLM   STDERR T PRT MAXDEC=2 ;
VAR   age marital ;
```

Exhibit 8.11

Specifying Options for PROC MEANS (continued)

Output

```
Variable  Label                               N          Mean        Std Dev
------------------------------------------------------------------------------
AGE       Age of Respondent                  52         44.48          17.18
MARITAL   Respondent Marital Status          52          2.44           1.66
------------------------------------------------------------------------------

Variable  Label                                  Minimum       Maximum
------------------------------------------------------------------------
AGE       Age of Respondent                        19.00         78.00
MARITAL   Respondent Marital Status                 1.00          5.00
------------------------------------------------------------------------

Variable  Label                         Lower 95.0% CLM  Upper 95.0% CLM
-----------------------------------------------------------------------
AGE       Age of Respondent                       39.70            49.26
MARITAL   Respondent Marital Status                1.98             2.90
-----------------------------------------------------------------------

Variable  Label                              Std Error             T
--------------------------------------------------------------------
AGE       Age of Respondent                       2.38         18.67
MARITAL   Respondent Marital Status               0.23         10.60
--------------------------------------------------------------------

Variable  Label                              Prob>|T|
-----------------------------------------------------
AGE       Age of Respondent                    0.0001
MARITAL   Respondent Marital Status            0.0001
-----------------------------------------------------
```

PROC UNIVARIATE: Obtaining Detailed Statistics, Tables, and Graphs

Although PROC MEANS provides many basic summary statistics for continuous variables, it does not provide a detailed look at their distributions. PROC UNIVARIATE provides most of the descriptive and inferential statistics for central tendency and variability given by PROC MEANS. It also gives additional descriptive statistics such as the median, skewness, and kurtosis. It can optionally plot stem-and-leaf, box plot, and normal probability graphs. These show the shape of a variable's frequency distribution and any extreme values. Finally, it also provides tests of the hypothesis that the values of the variable found in the data at hand were drawn from a normal distribution.

Basic UNIVARIATE Syntax

A wide variety of statistics, quantile values, and extreme values may be obtained by running PROC UNIVARIATE with no options. Use a VAR statement to give the names of the variables for which you desire detailed statistics. Here, as for other procedures, a VAR statement is necessary to prevent calculation of statistics on all numeric variables in the input dataset.

An ID statement is optional, but very desirable for identifying extreme values when the dataset contains a clear case identification variable. Each observation in the listing of extreme values will be labeled with the value of the variable in the ID statement (this value is often a social security number or other identifier, such as last name). The copious output shown in Exhibit 8.12 was generated by a rather small amount of SAS code. It requests that a PROC UNIVARIATE be run on our now familiar `age` variable and uses the character region variable (`cregion`) as an identification variable.

Exhibit 8.12

A Basic PROC UNIVARIATE, Using an ID Statement

Program

```
LIBNAME data 'location' ;
PROC UNIVARIATE    DATA=data.gss1991;
VAR    age      ;
ID     cregion ;
RUN;
```

Output

```
Variable=AGE                Age of Respondent

                          Moments

 N                  52   Sum Wgts          52
 Mean         44.48077   Sum             2313
 Std Dev       17.1824   Variance    295.2349
 Skewness     0.416267   Kurtosis    -0.97732
 USS            117941   CSS         15056.98
 CV           38.62883   Std Mean     2.38277
 T:Mean=0     18.66767   Pr>|T|        0.0001
 Num ^= 0           52   Num > 0           52
 M(Sign)            26   Pr>=|M|       0.0001
 Sgn Rank          689   Pr>=|S|       0.0001

                    Quantiles(Def=5)

 100% Max           78          99%        78
  75% Q3            58          95%        74
  50% Med         39.5          90%        71
  25% Q1          32.5          10%        23
   0% Min           19           5%        20
                                 1%        19
 Range              59
 Q3-Q1            25.5
 Mode               36

                         Extremes

    Lowest     ID          Highest     ID
        19(E. N. Ce)           71(Pacific )
        20(W. N. Ce)           74(E. N. Ce)
        20(Mountain)           74(Mid. Atl)
        21(Pacific )           77(Pacific )
        22(Sou. Atl)           78(E. S. Ce)
```

As you can see in Exhibit 8.12, a great deal of information about a variable is packed into a small area. There are three sections in PROC UNIVARIATE's basic output. In order, they are labeled `Moments`, `Quantiles`, and `Extremes`. Exhibit 8.13 provides a brief description of each of the summary and inferential statistics in the `Moments` section.

Exhibit 8.13

Statistics in the Moments Portion of UNIVARIATE Output

Output Label	Description
N	Number of cases with no missing values
SumWgts	Sum of weights for variable on WEIGHT statement (equals N if no WEIGHT statement is used)
Mean	Arithmetic mean of variable values
Sum	Sum of variable values used for calculating mean
Std Dev	Sample standard deviation (square root of variance)
Variance	Sample variance
Skewness	Skewness (measure of distribution symmetry)
Kurtosis	Kurtosis (measure of distribution peakedness or flatness)
USS	Total of the squared value of each case
CSS	Total of squared value of difference between the mean and the value of each case
CV	Coefficient of variation
Std Mean	Standard error of the mean
T:Mean=0	Student's t for test of hypothesis that population mean equals zero
Prob>|T|	Two-tailed P value for H_0: Mean=0
Num ^=0	Number of nonmissing values not equal to 0
Num >0	Number of nonmissing values greater than 0
M(Sign)	Sign test of hypothesis that population median is equal to 0
Pr>|M|	Significance level for M
Sgn Rank	Value of sign rank test (S) that population median is equal to 0
Pr>|S|	Significance level for S

The quantiles of the distribution of the variable are detailed in the section headed `Quantiles`. The quantile labeled `25% Q1` in the output is the value that includes the lowest-valued 25% of the cases. It's also referred to as the first quartile, because it includes the first fourth of the cases. The quantile labeled `50% Med.` is the familiar median value — it separates the higher-valued 50% of the cases from the lower-valued 50%. PROC UNIVARIATE is the only basic SAS procedure that calculates the median. The upper quartile is labeled `75% Q3`. The statistics given below the quantiles are the range (maximum value minus minimum value), the interquartile range, labeled `Q3-Q1` (third quartile minus first quartile), and the mode, or most frequent value.

There are a variety of appropriate methods for calculating quantiles of distributions. The `Def=5` section heading indicates that UNIVARIATE used the fifth of its five methods in this case. The differences between methods are subtle and usually of no concern. However, the details are included in the SAS Institute documentation for

PROC UNIVARIATE. See "Navigating the Documentation Maze," in Chapter 7, for more details on documentation for this and other statistical PROCs.

The section of the output with the heading `Extremes` lists the five lowest-valued and the five highest-valued cases. If no ID statement is used, the sequential numbers of the observations in the dataset are printed in parentheses next to the values. When an ID statement is used, the first eight characters of the values of the variable named on the ID statement for those cases are placed within the parentheses. In this example, the identification variable is the U.S. Census Region name. Examination of extreme values can help detect both invalid variable values and outliers that may compromise more sophisticated analyses.

Display and Analytical Options

Three options may be placed on the PROC UNIVARIATE statement. These may request a stem-and-leaf plot, a box plot, and a cumulative normal plot; a test for the normality of frequency distribution; and a detailed frequency distribution for each variable on the VAR statement.

Use the PLOT option to produce the frequency distribution plots. The stem-and-leaf and box plots resemble those introduced by Tukey and associated with the EDA (Exploratory Data Analysis) approach to statistical analysis. If the number of cases is too large to produce an interpretable stem-and-leaf plot, PROC UNIVARIATE will instead produce a horizontal histogram.

Add the NORMAL option to produce one of two tests for the null hypothesis that a variable's values were drawn from a normally distributed population. When the number of observations with no missing values is 2000 or less, PROC UNIVARIATE produces the Shapiro-Wilk test statistic and associated probability. These are labeled `W:Normal` and `Pr<W` on the procedure's output. If the number of nonmissing values for a variable is greater than 2000, PROC UNIVARIATE calculates the Kolmogorov statistic and associated probability. These are labeled `D:Normal` and `Pr>D`. One of these statistics will be the last entry in the Moments section of the output when the NORMAL option is used.

Specifying the FREQ option on the PROC UNIVARIATE statement will produce a frequency distribution for the values of each variable named in the VAR statement. The output is more compact than that generated by PROC FREQ. Remember, however, that variables with large numbers of values will cause PROC UNIVARIATE to produce voluminous output when this option is used.

In Exhibit 8.14, we present edited output for PROC UNIVARIATE using the NORMAL and PLOT options. The data are a random sample of 200 numbers taken from a normally distributed population. Only the Moments section, showing the location of the test for normality, and the plots are shown. Note that the last line of the Moments section contains the Shapiro-Wilk statistic and its level of significance.

Exhibit 8.14

Using the NORMAL and PLOT Options in UNIVARIATE

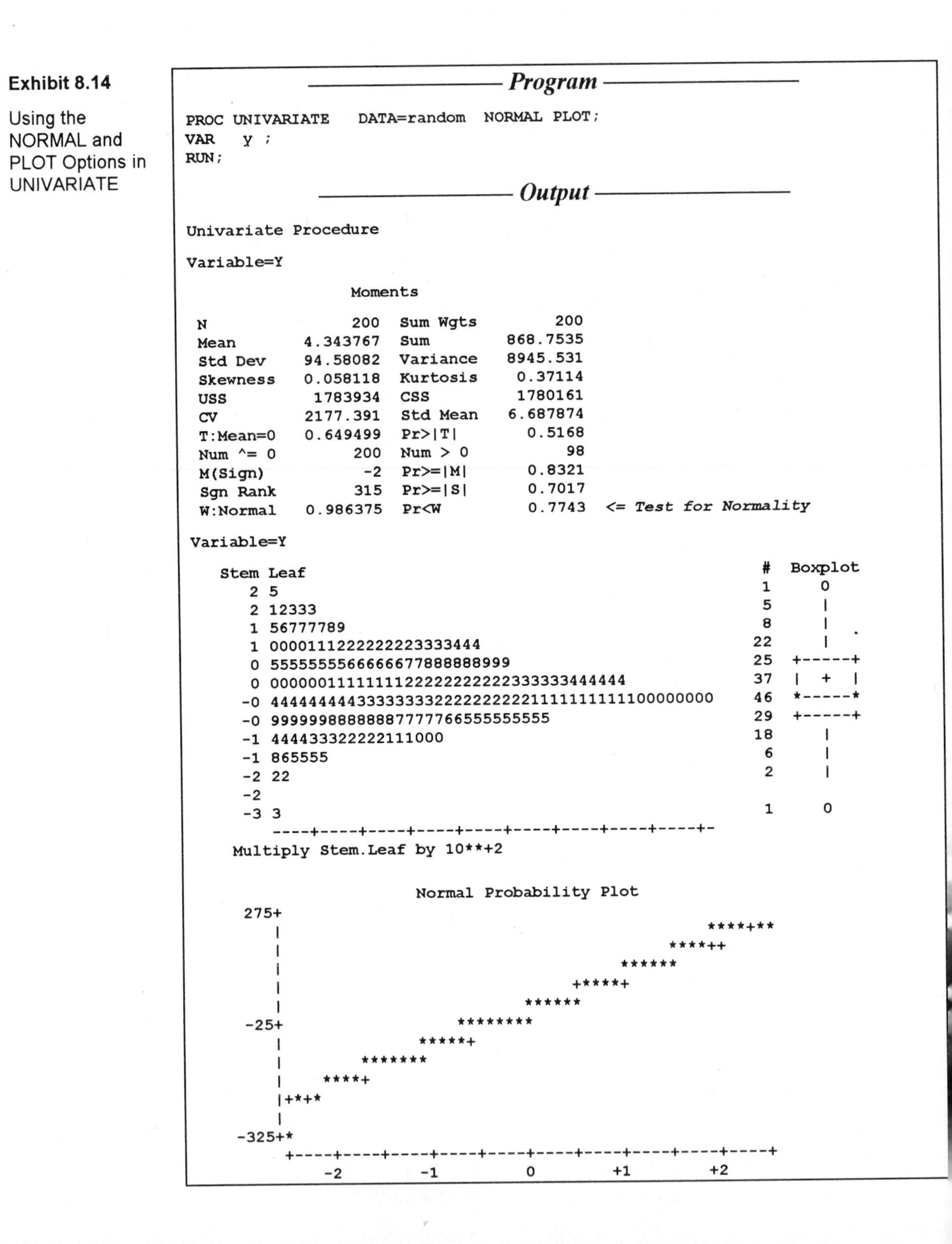

Program

```
PROC UNIVARIATE   DATA=random  NORMAL PLOT;
VAR   y ;
RUN;
```

Output

```
Univariate Procedure

Variable=Y

                  Moments

 N               200  Sum Wgts         200
 Mean       4.343767  Sum         868.7535
 Std Dev    94.58082  Variance    8945.531
 Skewness   0.058118  Kurtosis     0.37114
 USS         1783934  CSS          1780161
 CV         2177.391  Std Mean    6.687874
 T:Mean=0   0.649499  Pr>|T|        0.5168
 Num ^= 0        200  Num > 0           98
 M(Sign)          -2  Pr>=|M|       0.8321
 Sgn Rank        315  Pr>=|S|       0.7017
 W:Normal   0.986375  Pr<W          0.7743  <= Test for Normality

Variable=Y

   Stem Leaf                                               #  Boxplot
      2 5                                                  1     0
      2 12333                                              5     |
      1 56777789                                           8     |
      1 0000111222222223333444                            22     |
      0 5555555566666677888888999                         25  +-----+
      0 0000001111111222222222222333333444444             37  |  +  |
     -0 4444444443333333222222222211111111111000000000    46  *-----*
     -0 99999988888887777766555555555                     29  +-----+
     -1 444433322222111000                                18     |
     -1 865555                                             6     |
     -2 22                                                 2     |
     -2
     -3 3                                                  1     0
        ----+----+----+----+----+----+----+----+----+-
    Multiply Stem.Leaf by 10**+2

                    Normal Probability Plot
     275+
        |                                                   *****+**
        |                                               *****++
        |                                          ******
        |                                     +*****+
        |                                ******
    -25+                          ********
        |                      *****+
        |                *******
        |           ****+
        |+++*
        |
   -325+*
        +----+----+----+----+----+----+----+----+----+----+
            -2        -1         0        +1        +2
```

The stem-and-leaf plot, on the left of the first plot, is similar to a horizontal histogram in that the length of the bars, or "leaves," is based on the frequency of a range of values. Each bar, however, is made up of numbers representing the first few digits of the actual values of the variable. The first digits are the ones listed under `Stem` and the next are those listed under `Leaf`. If necessary, the lower-order digits are rounded. For this example variable (`Y`), the highest value was 249.8, which is represented with a stem of 2 and a leaf of 5 at the very top of the plot. The note at the bottom of the stem-and-leaf plot tells you that you must `Multiply Stem.Leaf by 10**+2` to approximate the actual magnitude of the data values represented by the plot. The single asterisk means "multiply by" and the double asterisk means "raise to the power of." 249.8, rounded up, is 2.5 * 10**+2, or 250.

The box plot to the right of the stem-and-leaf plot gives another impression of the frequency distribution. The box demarcated by the horizontal dashes flanked by + signs represents the interquartile range of values. The location of the median is indicated by the dashed line flanked by asterisks. The location of the mean is given by a single + sign within the box. The vertical lines, or "whiskers," indicate values that are within a distance of 1.5 times the interquartile range from the upper and lower quartiles. Values that are more extreme but within 3 times the interquartile range are represented by o's. Very extreme values, beyond 3 times the interquartile range, which were not present in this example data, are plotted with an asterisk. The vertical lines, + signs, zeros, and asterisks may extend from both the top and bottom of the box.

The graph under the heading `Normal Probability Plot`, produced by the PLOT option, compares the distribution of the values of a variable against a theoretically normal distribution of values with the same mean and standard deviation as those of the variable. Technically speaking, it is a quantile-quantile plot in which the quantile values of the variable run along the vertical axis and those of the theoretically normal variable run along the horizontal axis. The data points are marked with an asterisk. A diagonal reference line representing perfect normality is also plotted with + signs. The more the data depart from a normal distribution, the greater the number of + signs that are visible. Since our example data were drawn from a normally distributed population, very few + signs are visible in the plot in Exhibit 8.14.

RECAP

Syntax Summary

A summary of the syntax of the statistical procedures covered in this chapter follows. (See the syntax description in Chapter 1 for an explanation of the notation.)

PROC FREQ for generating frequency distributions:

```
PROC FREQ <DATA=input_dataset> <ORDER=FREQ>;
TABLES varname1 <varname2 ...> </>
      <MISSING> <NOCUM>;
```

PROC CHART for creating bar charts and histograms:

```
PROC CHART <DATA=input_dataset> ;
HBAR | VBAR varname1 varname2 ... </>
      <ASCENDING | DESCENDING>
      <DISCRETE> <LEVELS= nbars>
      <MIDPOINTS= val1 val2 <val3 ...>|
       MIDPOINTS= low TO high BY stepval>
      <MISSING> <NOSTATS>*
      <SPACE= nspaces> ;
```

* The NOSTATS option may be used only on an HBAR statement.

PROC MEANS for obtaining summary statistics:

```
PROC MEANS <DATA=input_dataset>  <ALPHA= plevel>
      <CLM> <MAXDEC= nplaces> <MEAN>
      <N> <NMISS> <PRT>
      <STD> <STDERR> <T>
      <VAR> <VARDEF=N>;
VAR varname1 <varname2 ...>;
```

PROC UNIVARIATE for getting detailed summary statistics and distribution plots:

```
PROC UNIVARIATE <DATA=input_dataset>
      <FREQ> <NORMAL> <PLOT> ;
VAR varname1 <varname2 ...> ;
ID idvarname ;
```

Content Review

In this chapter, we have given you an introduction to using four SAS procedures for obtaining tabular frequency distributions, summary statistics, and graphs of the distributions for individual variables — PROC FREQ, PROC CHART, PROC MEANS, and PROC UNIVARIATE. Examining the distributions of variables can be an end in itself, but is usually the first step in analyzing relationships among variables. We will provide you with an overview of SAS PROCs commonly employed to examine relationships among variables in the following chapters. Chapter 9 will deal with statistics for predicting continuous variables such as regression and analysis of variance. Chapter 10 will review techniques for predicting variables that are discrete or categorical in nature.

Statistics for Relationships with Continuous Dependent Variables

Researchers are often concerned with examining the relationships among two or more variables. They are often interested in seeing how well the values of one variable — usually called the dependent, or criterion, variable — can be predicted from the values of one or more independent variables or predictors.

The choice of statistical technique is determined in part by the measurement scale of the dependent and independent variables. When the dependent variable is an interval or ratio scale and the usual least squares assumptions can be made, the choice is generally governed by the measurement scales of the independent variables. When all the independent variables, or IVs, are categorical, ANOVA is generally used. If the IVs are all continuous, bivariate or multiple regression is usually employed. If some IVs are continuous and others are categorical, analysis of covariance (ANCOVA) is generally preferred. This chapter will

- ❑ Show you how to use PROC CHART, PROC UNIVARIATE, and PROC PLOT to visualize bivariate relationships with continuous dependent variables
- ❑ Introduce you to PROC TTEST and PROC GLM for performing ANOVA and related tests of differences of means
- ❑ Explain the use of PROC REG for performing basic regression analysis
- ❑ Review the use of PROC GLM and PROC REG for performing an analysis of covariance

ANOVA: Statistics for Categorical Independent Variables

In many situations, researchers compare the means of a continuous dependent variable across categories, or groups, of one or more independent variables. These categories divide observations into mutually exclusive groups. A difference among group means indicates a relationship between the categorical variable(s) defining the groups and the continuous dependent variable.

Groups may occur naturally, as in human gender or race. They may also result from the random assignment of observations to experimental conditions, as in medical or psychological research. In all cases, the basic statistical question is whether differences in group means observed in the data are likely to be real or to result from error introduced through sampling or random assignment.

If there is just one categorical, independent variable with two categories, this question can be answered with the well-known t test. If there is a single independent variable defining more than two groups, a one-way ANOVA is usually employed to test for differences. If two categorical independent variables are used to cross-classify the observations, a two-way ANOVA may be employed. More complicated or *n*-way research designs are also possible.

In addition to wanting to know whether some or all of the group means differ from each other, researchers sometimes wish to compare means for specific groups to see whether pairs or sets of means differ. In some situations, tests of mean differences are derived from research hypotheses stated prior to performing statistical tests. These are referred to as *planned* tests. In other situations, the researcher has no prior hypothesis but would like to get an idea of whether the mean differences observed in the data are likely to exist in the population. Such tests are usually referred to as *post hoc* or *unplanned.*

PROC TTEST, as its name implies, may be used to perform t tests of differences between two group means. PROC GLM can be used to perform one-way, two-way, and *n*-way ANOVAs. Before discussing these procedures, however, let's take a look at two procedures with a more visual approach to examining relationships between variables. PROC CHART and PROC UNIVARIATE may be used to display and compare group means and within-group frequency distributions.

Visualizing Group Distributions with PROC CHART

It is often useful to visually examine the distribution of the dependent variable within categories of the independent variable or variables. This is typically done prior to performing a t test or a one-way ANOVA. This practice can help you check for atypical data values, assess normality and equality of variance assumptions, and provide information on ranges of values.

PROC CHART can be used with the GROUP= option on the HBAR or VBAR statement. This produces a convenient bar chart for visually comparing the distributions of a continuous dependent variable for each category of the grouping variable. The GROUP= option is separated from the name of the dependent variable by a slash (/). The name of the variable used to categorize the observations follows the equal sign.

Exhibit 9.1 shows the program used to create the dataset used in several examples later in this chapter. The data read in in the DATA step are from Maxwell and Delaney (1990), page 303. The observed proportion of all play behavior with parents that involved pretending is given for male and female infants who were 7, 10, and 13 months old. A child psychologist might be interested in comparing the distributions of the variable `pretend` for each value of `age`.

Exhibit 9.1

Program to Create Pretend Behavior Dataset

```
LIBNAME    data   'location' ;

*** Build a permanent SAS dataset *** ;
DATA   data.baby ;
       INPUT   subject  age    gender $   pretend   ;

       LABEL   subject= "Infant's ID Number"
               age    = "Infant's Age in Months"
               gender = "Infant's Gender"
               pretend= "Pretend Behavior" ;

DATALINES;
  1       7      Female       .02
  2       7      Female       .01
  3       7      Female       .07
  4       7      Female       .04
  5       7      Female       .01
  6       7      Female       .09
  7       7      Female       .05
  8       7      Female       .06
  9       7      Male         .05
 10       7      Male         .01
 11       7      Male         .04
 12       7      Male         .03
 13       7      Male         .02
 14       7      Male         .02
 15       7      Male         .13
 16       7      Male         .06
 17      10      Female       .15
 18      10      Female       .11
 19      10      Female       .22
 20      10      Female       .05
 21      10      Female       .09
 22      10      Female       .05
 23      10      Female       .15
 24      10      Female       .11
 25      10      Male         .14
 26      10      Male         .21
 27      10      Male         .06
 28      10      Male         .12
```

Exhibit 9.1

Program to Create Pretend Behavior Dataset (continued)

```
29      10      Male        .11
30      10      Male        .19
31      10      Male        .12
32      10      Male        .04
33      13      Female      .09
34      13      Female      .03
35      13      Female      .18
36      13      Female      .12
37      13      Female      .18
38      13      Female      .43
39      13      Female      .24
40      13      Female      .40
41      13      Male        .02
42      13      Male        .19
43      13      Male        .15
44      13      Male        .07
45      13      Male        .45
46      13      Male        .20
47      13      Male        .49
48      13      Male        .19
```

Exhibit 9.2 uses PROC CHART to compare data for age groups.

Exhibit 9.2

PROC CHART with GROUP= Option

Program

```
*** Create a bar chart that compares age groups ***;
PROC CHART  DATA=data.baby;
     HBAR   pretend  / GROUP=age ;
TITLE1 'Distribution of' ;
TITLE2 'Proportion of Pretend Behavior' ;
TITLE3 'for Each Age Group' ;
RUN;
```

Output

```
Distribution of
Proportion of Pretend Behavior
for Each Age Group

AGE    PRETEND                                 Cum.              Cum.
       Midpoint                         Freq   Freq   Percent   Percent
                 |
  7       0.00   |**************           7      7     14.58     14.58
          0.08   |****************         8     15     16.67     31.25
          0.16   |**                       1     16      2.08     33.33
          0.24   |                         0     16      0.00     33.33
          0.32   |                         0     16      0.00     33.33
          0.40   |                         0     16      0.00     33.33
          0.48   |                         0     16      0.00     33.33
                 |
```

Exhibit 9.2

PROC CHART with GROUP= Option (continued)

```
10        0.00   |                    0    16     0.00    33.33
          0.08   |****************    8    24    16.67    50.00
          0.16   |************        6    30    12.50    62.50
          0.24   |****                2    32     4.17    66.67
          0.32   |                    0    32     0.00    66.67
          0.40   |                    0    32     0.00    66.67
          0.48   |                    0    32     0.00    66.67
                 |
13        0.00   |****                2    34     4.17    70.83
          0.08   |****                2    36     4.17    75.00
          0.16   |************        6    42    12.50    87.50
          0.24   |****                2    44     4.17    91.67
          0.32   |                    0    44     0.00    91.67
          0.40   |****                2    46     4.17    95.83
          0.48   |****                2    48     4.17   100.00
                 |
                 ----+---+---+---+
                     2   4   6   8

                     Frequency
```

Visualizing Distributions with PROC UNIVARIATE

PROC CHART can provide a pretty good picture of the frequency distributions of a continuous variable within categories of a grouping variable. It cannot, however, clearly show the location of group means or the degree of their variability. PROC UNIVARIATE can produce side-by-side box plots that do both. Simply add a PLOT option to the PROC UNIVARIATE statement and use a BY statement to identify the grouping variable(s).

PROC SORT, discussed in Chapter 6, may be used to sort the dataset before PROC UNIVARIATE is invoked. As you can see from Exhibit 9.2, the data used here are already ordered by age, so, strictly speaking, PROC SORT is not needed in this example. In most situations, however, the data are not already in the correct order and the sort is required.

The last page of the PROC UNIVARIATE output contains the side-by-side box plots. These plots are shown in Exhibit 9.3. You can easily see that both the mean and the variability of proportion of pretend play seem to increase with age.

Exhibit 9.3

Side-by-Side Box Plots from PROC UNIVARIATE

Program

```
LIBNAME data  'location' ;

/* Sort data by variable to be used in */
/* PROC UNIVARIATE BY statement        */
PROC SORT    DATA=data.baby    OUT=temp ;
     BY  age ;
RUN;
```

Exhibit 9.3

Side-by-Side Box Plots from PROC UNIVARIATE (continued)

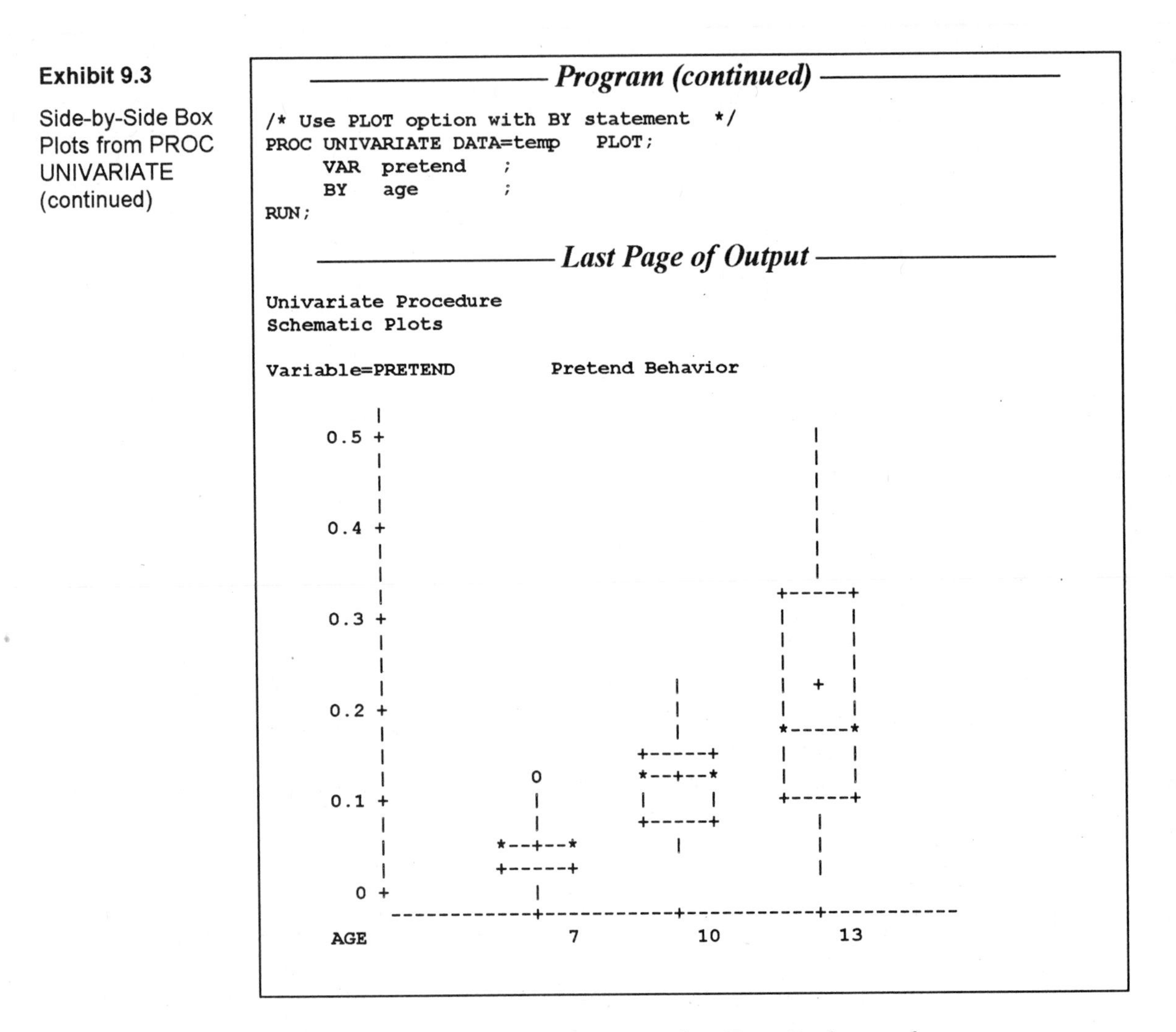

Program (continued)

```
/* Use PLOT option with BY statement  */
PROC UNIVARIATE DATA=temp   PLOT;
     VAR  pretend   ;
     BY   age       ;
RUN;
```

Last Page of Output

```
Univariate Procedure
Schematic Plots

Variable=PRETEND            Pretend Behavior

          |
      0.5 +                                          |
          |                                          |
          |                                          |
          |                                          |
      0.4 +                                          |
          |                                          |
          |                                          |
          |                                       +-----+
      0.3 +                                       |     |
          |                                       |     |
          |                                       |     |
          |                          |            |  +  |
      0.2 +                          |            |     |
          |                          |            *-----*
          |                       +-----+         |     |
          |              0        *--+--*         |     |
      0.1 +              |        |     |         +-----+
          |              |        +-----+            |
          |           *--+--*        |               |
          |           +-----+                        |
        0 +              |
           ------------+-----------+-----------+-----------
  AGE                  7          10          13
```

Testing Differences in Means for Two Independent Groups: PROC TTEST

PROC TTEST furnishes you with the conventional t test of significance for difference of means. It assumes the dependent variables are continuous and are taken from two independent samples. It calculates the exact t test assuming equal within-group variances in the population. It also computes the approximate t test based on the assumption that the variances are not equal. TTEST also provides a specialized F test of the null hypothesis of equal population variances.

The code in Exhibit 9.4 illustrates the basic syntax for PROC TTEST. The CLASS statement is required. It must include the name of a single numeric or character SAS variable whose values divide the observations in the data into two groups of non-missing values. In this case, we're testing for a difference between male and female infants, so we use the variable `gender` on the CLASS statement. The VAR statement should contain the name of at least one continuous dependent variable. Separate analyses will be performed for each variable named on the VAR statement. No adjustment is made to the p-values for t tests when tests for multiple variables are analyzed.

Exhibit 9.4

PROC TTEST

Program

```
LIBNAME  DATA  'location' ;

PROC TTEST  DATA=data.baby ;
     CLASS  gender  ;
     VAR    pretend ;
TITLE1  'Difference Between Male and Female';
TITLE2  'Infant Pretending Behavior'         ;
RUN;
```

Output

```
Difference Between Male and Female
Infant Pretending Behavior

TTEST PROCEDURE

Variable: PRETEND        Pretend Behavior

GENDER        N            Mean           Std Dev         Std Error
-------------------------------------------------------------------
Female       24      0.12291667        0.11035434        0.02252598
Male         24      0.12958333        0.12369244        0.02524861

Variances         T        DF     Prob>|T|
------------------------------------------
Unequal     -0.1970      45.4       0.8447
Equal       -0.1970      46.0       0.8447

For H0: Variances are equal, F' = 1.26   DF = (23,23)   Prob>F' = 0.5887
```

The output from the PROC TTEST in Exhibit 9.4 is easy to interpret. The name of the dependent variable (`pretend`) is displayed next to the heading `Variable:` at the top left of the output. The table immediately beneath this line of output gives descriptive statistics for the dependent variable for each of the two groups defined by the variable (`gender`) named on the CLASS statement. These statistics include the number of observations, labeled `N`, the mean of the continuous variable, labeled `Mean`, the standard deviation of each mean, labeled `Std Dev`, and the standard error of each mean, labeled `Std Error`.

The second table gives t values under the heading `T`, degrees of freedom, and the significance level of the t for both the standard t test (assuming equal group variances) and the test that assumes unequal group variances. The last line of the output gives the folded F statistic, along with its degrees of freedom and level of significance. Note that a statistically significant F statistic here means that a null hypothesis of equal group variances is rejected.

PROC TTEST may exclude an observation from the calculation of its descriptive and inferential statistics. This happens when a variable named on the VAR statement has a missing value or when the CLASS variable is missing for the observation. That is, TTEST uses only nonmissing pairs of CLASS and VAR statement variables. This treatment of missing values means that the number of observations in each group may vary for each variable on the VAR statement. The implications of this variation for particular research situations can range from inconsequential to severe.

Testing Differences in Means for Several Independent Groups: PROC GLM

PROC GLM may be used for assessing differences in means when a grouping variable has more than two nonmissing values. PROC ANOVA may also be used for one-way ANOVA. However, PROC GLM is more generally applicable for analyses of more complex designs with unequal cell sizes. It also allows preplanned tests of differences of cell means. For this reason, we recommend that you *always* use PROC GLM for ANOVA.

GLM Syntax The example code in Exhibit 9.5 shows the basic syntax for using PROC GLM to perform a one-way ANOVA. You must use a CLASS statement that names a single categorical independent variable. It must precede a MODEL statement. You must also employ a MODEL statement that expresses the desired analysis in an equation-like manner. The name of the continuous dependent variable is placed to the left of an equal sign. The name of the categorical independent variable is placed to the right. There can be only one CLASS and one MODEL statement following each PROC GLM statement.

The MEANS statement is optional but very useful. It instructs PROC GLM to print the number of observations, the mean, and the standard deviation of the dependent variable for each of the groups defined by the independent variable. Unlike PROC TTEST, PROC GLM does not do this automatically. Be sure to place the name of the independent variable on the MEANS statement and include the statement after the MODEL statement.

In Exhibit 9.5, we test for differences in means among the age groups for pretending behavior. The SAS variable name of the grouping variable (`age`) is used in the CLASS statement, the MEANS statement, and the MODEL statement. The name of the dependent variable (`pretend`) is used only in the MODEL statement.

Exhibit 9.5

One-Way ANOVA with PROC GLM

Program

```
PROC GLM    DATA=data.baby ;

     /* Name of Independent Variable   */
     CLASS  age ;

     /* Name of dependent variable on left of = */
     /* Name of independent variable on right   */
     MODEL  pretend = age  ;

     /* Print means and SD's of depvar */
     MEANS  age ;

TITLE1  'One-Way ANOVA for Infant'    ;
TITLE2  'Pretending Behavior by Age' ;
RUN;
```

Output

```
One-Way ANOVA for Infant
Pretending Behavior by Age

General Linear Models Procedure
Class Level Information

Class    Levels    Values

AGE           3    7 10 13

Number of observations in data set = 48
```

New Page of Output

```
One-Way ANOVA for Infant
Pretending Behavior by Age

General Linear Models Procedure

Dependent Variable: PRETEND    Pct. Pretend Behavior
                                    Sum of          Mean
Source                   DF        Squares        Square  F Value     Pr > F

Model                     2     0.23213750    0.11606875    13.05     0.0001
Error                    45     0.40038750    0.00889750
Corrected Total          47     0.63252500

               R-Square           C.V.        Root MSE        PRETEND Mean
               0.367001       74.71411         0.09433             0.12625

Source                   DF    Type I SS   Mean Square  F Value     Pr > F
AGE                       2   0.23213750    0.11606875    13.05     0.0001

Source                   DF  Type III SS   Mean Square  F Value     Pr > F
AGE                       2   0.23213750    0.11606875    13.05     0.0001
```

Exhibit 9.5

One-Way ANOVA with PROC GLM (continued)

New Page of Output

```
One Way ANOVA for Infant
Pretending Behavior by Age

General Linear Models Procedure

Level of            -----------PRETEND-----------
AGE          N          Mean              SD

7           16      0.04437500       0.03285701
10          16      0.12000000       0.05549775
13          16      0.21437500       0.15010968
```

Interpreting GLM Output for One-Way ANOVA As you can see in Exhibit 9.5, the output for PROC GLM provides basic ANOVA statistics and also some additional measures of the fit of the linear model for the ANOVA to the data. The first section of the output lists the name of the independent variable (`age`), the number of levels or groups it has, and the numeric or character code for each level under the heading `Class Level Information`. The number of observations eligible for the analysis appears after the `Class Level Information` heading. If either the dependent or the independent variable on an observation has a missing value, PROC GLM will not include that observation in the analysis. It will also print a note about the number of observations actually included in the analysis. Since these data have no missing values, a note does not appear.

The next section of the output begins with the name of the dependent variable (`pretend`) specified on the MODEL statement appearing at the top left of the output, just to the right of the label `Dependent Variable:`. Beneath this line is an ANOVA table that gives the explained sum of squares and associated statistics in the row labeled `Model`, the error sum of squares in the row labeled `Error`, and the total sum of squares in the row labeled `Corrected Total`. For one-way ANOVAs, the `F Value` and the associated level of significance, labeled `Pr > F`, for the `Model` sum of squares is the test of the null hypothesis of equality of the group means. Just below the ANOVA table, you will find some measures of the fit of the linear model for the ANOVA to the data. These include `R-Square`, which is also the proportion of the total sum of squares explained by the model.

For a one-way ANOVA, the explained sum of squares is due to only one variable or factor. Thus, the information on the sources of variation in the model located below the measures of fit is identical to that in the `Model` row of the ANOVA table. The difference between the sums of squares labeled `Type I SS` and those labeled `Type III SS` will be dealt with later in this chapter, when *n*-way ANOVA is discussed. There is no difference between them for one-way ANOVAs.

The last page of the output gives the means and standard deviations of the dependent variable for each level of the CLASS variable. This output was generated by the MEANS statement.

A significant F statistic for the model sum of squares lends support to the hypothesis that some of the group means differ from others. It does not, however, compare specific group means. In some research situations, the investigator has no prior hypotheses about possible mean differences but would still like to know which group means differ in a statistically significant manner.

Specifying Post Hoc Comparisons

There are many statistical techniques for performing such post hoc, or unplanned, multiple comparisons. Most of them may be obtained by using options on the MEANS statement. Some of the more commonly used techniques and their associated option names are summarized in Exhibit 9.6. For complete details on all available techniques, consult the documentation for PROC GLM cited in the Statistical Roadmap in Chapter 7 or see Littell, Freund, and Spector (1991). For a useful discussion of many of the techniques available in PROC GLM and how to choose among them, see Maxwell and Delaney (1990), Chapter 5.

Exhibit 9.6

Multiple Comparison Methods in PROC GLM

Option Name	Technique
BON	Bonferonni t tests of pairs of means
DUNCAN	Duncan multiple range test
SCHEFFE	Scheffe's multiple comparison technique
SNK	Student-Newman-Keuls multiple range test
LSD	Least significant difference pairwise t tests
TUKEY	Tukey's studentized range test

Specifying Planned Comparions: The CONTRAST Statement In many research situations, the investigator has thought about what means to compare prior to the analysis. These are *planned comparisons*, and are performed with one or more CONTRAST statements. CONTRAST statements must come after a MODEL statement. They define a linear combination of means to be tested. For example, if an analysis involves an independent variable named `method` with four groups coded A, B, C, and D, the following CONTRAST statements would compare the mean for group A with that for group C and the average of groups B and C with the mean of group D:

```
CONTRAST 'A vs. D'       method  1  0   -1 0;
CONTRAST 'B & C vs. D'   method  0 .5   .5 -1;
```

The text within quotes following `CONTRAST` is an optional label you may supply for the contrast. It appears in the output and is useful for identifying output related to a particular contrast. Labels longer than 20 characters are shortened to 20 characters in the output. The label is followed by the name of the variable used in the CLASS statement. The numbers that follow define the linear combination of the means to be tested. For example, the first contrast, above, would be for the null hypothesis that

$$1*MEAN_A + 0*MEAN_B + -1*MEAN_C + 0*MEAN_D = 0.$$

The coefficient order corresponds to the alphabetical or numerical order of the values of the independent variable as listed in the `Class Levels Information` section of the output.

Details about the formulation of linear contrasts in ANOVA can be found in most statistics texts covering ANOVA and experimental design, such as Maxwell and Delaney (1990). A well-written explanation of testing linear contrasts using the CONTRAST statement in GLM is provided in Littell, Freund, and Spector (1991), pages 68–74 and 95–104. Remember that although PROC GLM will execute many CONTRAST statements, it does not check on their independence or control the experimentwise error rate.

You would not usually combine planned and unplanned contrasts in one analysis. For brevity's sake, the code in Exhibit 9.7 illustrates how to do both. The option names for post hoc Bonferonni and LSD comparisons appear on the MEANS statement. Two CONTRAST statements are also included. The first CONTRAST statement tests the difference in means between the 10-month-old infants and the 13-month-olds. The second tests the difference between the means of the 10-month-olds and the 13-month-olds and the mean of the 7-month-olds.

The statistics on the first two pages of the output for this example are identical to those in Exhibit 9.5, so only the output for the post hoc and planned tests of mean differences is shown in Exhibit 9.7.

Exhibit 9.7

Planned and Post Hoc Comparisons of Means in PROC GLM

Program

```
PROC GLM    DATA=data.baby ;
 CLASS   age ;
 MODEL   pretend = age ;

 MEANS   age  /  BON   LSD ;

      CONTRAST '10 versus 13 Months'
                 age  0    1   -1 ;

      CONTRAST '7 versus 10 & 13 Months'
                 age  1   -.5   -.5 ;
TITLE1   'One-Way ANOVA for Infant Pretending Behavior' ;
TITLE2   'Unplanned and Preplanned Tests of Differences' ;
RUN;
```

Exhibit 9.7

Planned and Post Hoc Comparisons of Means in PROC GLM (continued)

Output

```
One-Way ANOVA for Infant Pretending Behavior
Unplanned and Preplanned Tests of Differences

General Linear Models Procedure

T tests (LSD) for variable: PRETEND

NOTE: This test controls the type I comparisonwise error rate not the
      experimentwise error rate.

Alpha= 0.05  df= 45  MSE= 0.008897
Critical Value of T= 2.01
Least Significant Difference= 0.0672

Means with the same letter are not significantly different.

             T Grouping             Mean       N  AGE

                      A          0.21438      16  13

                      B          0.12000      16  10

                      C          0.04438      16  7
```

New Page of Output

```
One-Way ANOVA for Infant Pretending Behavior
Unplanned and Preplanned Tests of Differences

General Linear Models Procedure

Bonferroni (Dunn) T tests for variable: PRETEND

NOTE: This test controls the type I experimentwise error rate, but
      generally has a higher type II error rate than REGWQ.

Alpha= 0.05  df= 45  MSE= 0.008897
Critical Value of T= 2.49
Minimum Significant Difference= 0.0829

Means with the same letter are not significantly different.

         Bon Grouping             Mean       N  AGE

                    A          0.21438      16  13

                    B          0.12000      16  10
                    B
                    B          0.04438      16  7
```

Exhibit 9.7

Planned and Post Hoc Comparisons of Means in PROC GLM (continued)

New Page of Output

```
One-Way ANOVA for Infant Pretending Behavior
nplanned and Preplanned Tests of Differences

General Linear Models Procedure

Dependent Variable: PRETEND    Pretend Behavior

Contrast                   DF    Contrast SS    Mean Square  F Value     Pr > F

10 versus 13 Months         1     0.07125313     0.07125313     8.01     0.0069
10 & 13 versus 7 Mon        1     0.16088438     0.16088438    18.08     0.0001
```

The output for each post hoc comparison technique is clearly labeled on a separate page. For these comparisons, PROC GLM uses the letters A, B, C, and so on to label groups of means that are *not* significantly different according to a particular test (e.g., Bonferonni, least significant differences, and so on). Thus, the LSD test in this instance shows each mean different from the rest and uses the labels A, B, and C. However, the Bonferonni test finds that the 13-month-olds differ from the 7- and 10-month-olds, but the latter do not differ from each other. The 13-month-old mean, therefore, is labeled with an A, whereas the other two groups are labeled with a B.

The output for all planned comparisons specified on CONTRAST statements is condensed into one table. For this example, the table is located on the last page of the output. Each row in the table provides the result for a CONTRAST statement. The label supplied in the statement is listed in the leftmost column. The contrast degrees of freedom, sum of squares, mean squared error, F value, and level of significance are supplied for each contrast.

Using PROC GLM for Two-Way to N-Way ANOVAs

When there is more than one independent grouping variable, or factor, researchers are usually interested not only in simple differences in means but also in partitioning the explained sums of squares. This partitioning identifies quantities attributable to the main effects of individual variables and the interactions among them. Tests of significance of these main and interaction effects indicate whether variables have a relationship with the dependent variable, controlling for the effects of some or all of the other independent variables.

PROC GLM can be used for such analyses and is quite flexible in the types of models it can estimate and test. It would be impossible in this book to cover all the possible models and types of experimental designs that PROC GLM can handle. Therefore, we'll confine ourselves to explaining a two-way ANOVA for a complete factorial design. Readers are referred to Littell, Freund, and Spector (1991), Chapters 4 and 5, for a complete treatment of the capabilities of PROC GLM.

GLM Syntax for *N*-Way ANOVA The SAS program in Exhibit 9.8 illustrates the use of PROC GLM for an ANOVA involving two independent variables or factors. The CLASS statement serves the same purpose as in the one-way ANOVA example above. It identifies the categorical independent variables.

Exhibit 9.8

PROC GLM Output for a Two-Way ANOVA

Program

```
LIBNAME  data 'location' ;

PROC GLM     DATA=data.baby ;
     /* Age and gender are IVs */
     CLASS  age  gender    ;

     /* Model includes main and interaction effects */
     MODEL  pretend = age  gender  age*gender ;

     /* Get means for all three effects. */
     MEANS  age  gender  age*gender ;

TITLE1  'Two-Way ANOVA for Infant Pretending Behavior'   ;
TITLE2  'Main and Interaction Effects of Age and Gender'  ;
RUN;
```

Output

```
Two-Way ANOVA for Infant Pretending Behavior
Main and Interaction Effects of Age and Gender

General Linear Models Procedure
Class Level Information

Class     Levels    Values

AGE            3    7 10 13
GENDER         2    Female Male


Number of observations in data set = 48
```

New Page of Output

```
Two-Way ANOVA for Infant Pretending Behavior
Main and Interaction Effects of Age and Gender

General Linear Models Procedure

Dependent Variable: PRETEND   Pretend Behavior
                                    Sum of           Mean
Source                     DF      Squares         Square  F Value     Pr
             > F
Model                       5   0.23287500     0.04657500     4.89
             0.0013
Error                      42   0.39965000     0.00951548
Corrected Total            47   0.63252500
```

Exhibit 9.8

PROC GLM Output for a Two-Way ANOVA (continued)

R-Square	C.V.	Root MSE	PRETEND Mean
0.368167	77.26519	0.09755	0.12625

Source	DF	Type I SS	Mean Square	F Value	Pr > F
AGE	2	0.23213750	0.11606875	12.20	0.0001
GENDER	1	0.00053333	0.00053333	0.06	0.8140
AGE*GENDER	2	0.00020417	0.00010208	0.01	0.9893

Source	DF	Type III SS	Mean Square	F Value	Pr > F
AGE	2	0.23213750	0.11606875	12.20	0.0001
GENDER	1	0.00053333	0.00053333	0.06	0.8140
AGE*GENDER	2	0.00020417	0.00010208	0.01	0.9893

New Page of Output

Two-Way ANOVA for Infant Pretending Behavior
Main and Interaction Effects of Age and Gender

General Linear Models Procedure

Level of		-----------PRETEND-----------	
AGE	N	Mean	SD
7	16	0.04437500	0.03285701
10	16	0.12000000	0.05549775
13	16	0.21437500	0.15010968

Level of		-----------PRETEND-----------	
GENDER	N	Mean	SD
Female	24	0.12291667	0.11035434
Male	24	0.12958333	0.12369244

Level of	Level of		-----------PRETEND-----------	
AGE	GENDER	N	Mean	SD
7	Female	8	0.04375000	0.02924649
7	Male	8	0.04500000	0.03817254
10	Female	8	0.11625000	0.05680480
10	Male	8	0.12375000	0.05780200
13	Female	8	0.20875000	0.14247180
13	Male	8	0.22000000	0.16707569

The most important thing to notice on the MODEL statement in Exhibit 9.8 is the syntax for specifying the linear model effects used for the ANOVA — names of independent variables appearing alone specify main effects, whereas names separated by asterisks represent interaction or crossed effects. The linear model specified on the MODEL statement in this example lists each possible main and interaction effect for a two-way ANOVA. There are two main effects (`age` and `gender`) and one interaction effect (`age*gender`).

PROC GLM provides a shortcut for automatically generating both main and interaction effects. The MODEL statement above could have been written

```
MODEL  pretend =  age|gender ;
```

PROC GLM's MODEL statement also provides a syntax for specifying nested effects in which the nested variable precedes the nesting variable, which is enclosed in parentheses, as in

```
MODEL  pretend = age(gender) ;
```

This notation would be read as "age nested within gender." All of the above notations for specifying effects may be used in appropriate combinations to specify a model on the MODEL statement.

The MEANS statement in this example requests the main effect means for `age` and `gender` as well as those for each combination of the levels of `age` and `gender`. You may ask for the means associated with any effect specified on the MODEL statement. You cannot get means for any effect not specified on the MODEL statement. The MEANS statement is not required, but it is very useful.

Type I Versus Type III Sums of Squares The output shown in Exhibit 9.8 is similar to that of the one-way analysis earlier in this chapter. The primary difference is that now there are three sources of variation for the MODEL sums of squares, listed under the heading `Source` in the `Type I SS` and `Type III SS` sections. They correspond to each of the effects specified on the MODEL statement.

Because this example has an equal number of observations (eight) in each cell of the design, the design is balanced and the sums of squares for each effect are orthogonal. Thus, the results for both the Type I SS and the Type III SS are identical. When there are unequal numbers of observations in each cell of a design, these results will not be identical. Therefore, knowing the difference between the hypotheses tested by Type I and Type III sums of squares for unbalanced designs is important.

Type I sums of squares may be used for sequential or ordered tests of significance of the effects in the model. The sequential order in which the effects are tested is based on the order effects as specified on the MODEL statement, from left to right. For the example above, the test for the `age` main effect is uncontrolled for any other effects. The `gender` effect test is controlled only for the `age` main effect. The `age*gender` interaction test is controlled for the `age` and `gender` main effects. In other words, the Type I sum of squares test for an effect controls for all effects appearing *above* it in the list of effects, but *none* of the effects listed *below* it. The sum of the Type I sums of squares is, therefore, equal to the model or explained sum of squares. Remember that in unbalanced designs, the level of significance of the test for any effect based on a Type I sum of squares depends on the ordering of the tests.

Type III sums of squares are sometimes referred to as the *regression*, or *partial*, sums of squares in that the test for each effect is controlled for all effects listed *above* it *and below* it. Remember that in general, it is only in the case of balanced designs that the

Type III sums of squares are truly orthogonal. That is why for unbalanced designs the sum of the Type III sums of squares does not equal the model sum of squares.

It is possible to do unplanned multiple comparisons of means and planned contrasts of means for two-way ANOVAs in PROC GLM. However, doing so requires a detailed knowledge of ANOVA and of the appropriateness of various approaches for situations where interaction effects are or are not present. Therefore, we do not deal with this advanced topic here. Littell, Freund, and Spector (1991) is an invaluable reference for those who need further information on this topic.

Testing for Differences of Related Means

There are many research situations where one is interested in testing differences between means of one or more variables measured on the same group of observations across two or more points in time. This is often the case for child development or education researchers who collect data on groups of children over time. A common research design using this type of analysis is the pre-test, treatment, post-test research design. In these designs, some attribute of interest is measured on the same subjects prior to an experimental treatment and then again after it.

If only two means need to be compared, the code in Exhibit 9.9 shows how to obtain the familiar t test for comparing two means from related samples. This test is also referred to as the *paired t test* or *related t test*. In this example, we have modified the data for infant pretend behavior to make it appear that it was collected across time on the same babies. Note that each subject's proportions of pretend behavior at 7, 10, and 13 months of age are placed on the same data line but are given different SAS variable names (`month7, month10, month13`) on the INPUT statement. This illustrates the difference between the ways data for conventional experimental designs and for repeated-measures designs are recorded.

Let's say we are interested in testing the difference of means in pretend behavior between month 7 and month 13. The related t test is equivalent to testing whether the mean of the difference between two scores is equal to zero. Therefore, we calculate the difference between `month13` and `month7` in the DATA step below and assign it to the new variable `m13_m7`.

We place this new variable on the VAR statement of the second PROC MEANS statement. The options N, MEAN, STDERR, T, and PRT are needed on the second PROC MEANS statement to get all the information needed for the test of the hypothesis that the mean difference is zero, including the t statistic and its two-tailed probability level. The first PROC MEANS is not needed for the hypothesis test, but is useful because it provides the means of `month7` and `month13`. The output from the two PROC MEANS is shown in Exhibit 9.9.

Exhibit 9.9

Using PROC MEANS for Testing the Difference Between Two Related Means

Program

```
DATA babytime ;

         /* Note that measures for each time period are on */
         /* a single data line for each subject            */
          INPUT  subject $  gender $   month7  month10   month13;

     m13_m7 = month13 - month7 ;

     LABEL  gender   = 'Infant Gender'
            month7   = 'Pretend Behavior @  7 Months'
            month10  = 'Pretend Behavior @ 10 Months'
            month13  = 'Pretend Behavior @ 13 Months'
            m13_m7   = 'Difference between Month 13 & Month 7' ;
DATALINES;
    1      Female    0.02     0.15     0.09
    2      Female    0.01     0.11     0.03
    3      Female    0.07     0.22     0.18
    4      Female    0.04     0.05     0.12
    5      Female    0.01     0.09     0.18
    6      Female    0.09     0.05     0.43
    7      Female    0.05     0.15     0.24
    8      Female    0.06     0.11     0.40
    9      Male      0.05     0.14     0.02
   10      Male      0.01     0.21     0.19
   11      Male      0.04     0.06     0.15
   12      Male      0.03     0.12     0.07
   13      Male      0.02     0.11     0.45
   14      Male      0.02     0.19     0.20
   15      Male      0.13     0.12     0.49
   16      Male      0.06     0.04     0.19
RUN;

/* Get means for months 7 and 13. This is useful but not required. */
PROC MEANS DATA=babytime;
     VAR  month7 month13 ;
TITLE   'Proportion Pretend Behavior' ;
RUN;

/* Test difference between months 13 and 7 */
PROC MEANS N MEAN STDERR T PRT;
     VAR  m13_m7 ;
TITLE    'Two-Tailed Test for Difference between' ;
TITLE2   'Month 13 and Month 7' ;
RUN;
```

Exhibit 9.9

Using PROC MEANS for Testing the Difference Between Two Related Means (continued)

Output

```
Proportion Pretend Behavior

Variable  Label                                  N          Mean          Std Dev
------------------------------------------------------------------------------
MONTH7    Pretend Behavior @  7 Months  16     0.0443750     0.0328570
MONTH13   Pretend Behavior @ 13 Months  16     0.2143750     0.1501097
------------------------------------------------------------------------------

Variable  Label                              Minimum         Maximum
---------------------------------------------------------------------
MONTH7    Pretend Behavior @  7 Months     0.0100000       0.1300000
MONTH13   Pretend Behavior @ 13 Months     0.0200000       0.4900000
---------------------------------------------------------------------
```

New Page of Output

```
Two-Tailed Test for Difference between
Month 13 and Month 7

Analysis Variable : M13_M7 Difference between Month 13 & Month 7


 N          Mean       Std Error             T    Prob>|T|
-----------------------------------------------------------
16     0.1700000     0.0334415     5.0835054     0.0001
-----------------------------------------------------------
```

Repeated-Measures ANOVA If tests of the differences between more than two measurements taken over time are needed, a *repeated-measures ANOVA* is usually necessary. Such analyses are also referred to as *within-subjects designs*. The multivariate setup for this type of analysis using PROC GLM is illustrated in Exhibit 9.10. (Other approaches to the analysis of repeated measures designs are possible. See Maxwell and Delaney (1990) for a discussion of alternatives.)

We assume that the SAS dataset `babytime`, created in Exhibit 9.8, is available as a temporary dataset. Remember that in setting up the data for a repeated measures design, the dependent variable measured across time is given a different SAS variable name in the SAS dataset for each time period. In this example, we use the names `month7`, `month10`, and `month13`.

The first thing to notice in this example is that `month7`, `month10`, and `month13` are considered to be jointly dependent and are, therefore, listed to the left of the equal sign on the MODEL statement. Since there are no independent variables involved in this analysis, there are no variable names to the right of the equal sign. The NOUNI option on the MODEL statement suppresses some unnecessary output.

The REPEATED statement provides a SAS name of your choice for the factor represented by the dependent variables on the MODEL statement. This name is used to label the repeated-measures or within-subjects effect in the procedure output. In this example, the label is `age`. The NOU option suppresses more unnecessary output. The SUMMARY option provides the results of significance tests that make pairwise comparisons of the mean of the last time period (i.e., `month13`) with the means of each of the prior two time periods. Other patterns of comparisons are possible and are detailed in the PROC GLM documentation for the REPEATED statement cited in Chapter 7.

Exhibit 9.10

Repeated-Measures ANOVA

Program

```
PROC GLM  DATA=babytime ;
     /* Variables for each time period on left of = sign */
     MODEL  month7  month10  month13  =   /  NOUNI ;

     /* Name the repeated measures factor */
     REPEATED   age / NOU SUMMARY ;

TITLE1  'Repeated Measures Test of' ;
TITLE2  'Differences in Pretend Behavior' ;
RUN;
```

Output

```
Repeated Measures Test of
Differences in Pretend Behavior

General Linear Models Procedure

Number of observations in data set = 16

Repeated Measures ANOVA
Repeated Measures Level Information

Dependent Variable      MONTH7  MONTH10  MONTH13

      Level of AGE           1        2        3


Manova Test Criteria and Exact F Statistics for
the Hypothesis of no AGE Effect
H = Type III SS&CP Matrix for AGE   E = Error SS&CP Matrix

S=1     M=0     N=6

 Statistic                   Value          F      Num DF    Den DF  Pr > F

Wilks' Lambda              0.19480961    28.9325        2        14  0.0001
Pillai's Trace             0.80519039    28.9325        2        14  0.0001
Hotelling-Lawley Trace     4.13321696    28.9325        2        14  0.0001
Roy's Greatest Root        4.13321696    28.9325        2        14  0.0001
```

Exhibit 9.10

Repeated-Measures ANOVA (continued)

New Page of Output

```
Repeated Measures Test of
Differences in Pretend Behavior

General Linear Models Procedure
Repeated Measures ANOVA
ANOVA of Contrast Variables

AGE.N represents the contrast between the nth level of AGE and the last

Contrast Variable: AGE.1

Source                DF    Type III SS    Mean Square   F Value     Pr > F
MEAN                   1     0.46240000     0.46240000     25.84     0.0001
Error                 15     0.26840000     0.01789333

Contrast Variable: AGE.2

Source                DF    Type III SS    Mean Square   F Value     Pr > F
MEAN                   1     0.14250625     0.14250625      5.10     0.0393
Error                 15     0.41919375     0.02794625
```

The first page of the output gives the number of observations in the dataset. If there are missing values in the variables, the procedure reports the number of observations actually included in the repeated-measures analysis. If an observation has a missing value for any of the variables, it is excluded from the analysis.

The section labeled `Repeated Measures Level Information` shows how the variable names on the MODEL statement are assigned to levels of the repeated-measures factor named on the REPEATED statement. Here the variable `month7` is assigned to the first level, `month10` to the second, and `month13` to the third.

Following this section, under the heading `Manova Test Criteria and Exact F Statistics for the Hypothesis of no AGE Effect`, are multivariate F tests for the null hypothesis that all the means are equal. In this instance, that hypothesis is rejected by all tests at the .0001 level.

The next page gives the significance tests for the pairwise comparisons of the means of the first two ages with that of the last, under the heading `ANOVA of Contrast Variables`. The variable labeled `AGE.1` is the difference between `month7` and `month13`. The variable labeled `AGE.2` is the difference between `month10` and `month13`. For each variable, the sum of squares and F tests are given in the row of the table labeled `MEAN`.

Regression: Statistics for Continuous Independent Variables

Regression analysis is usually the preferred technique when the dependent variable and the independent variable(s) are continuous. The regression model contains estimates of the coefficients of a linear equation that predicts the values of the dependent variable from those of the independent variable(s). When only one independent variable is involved, the technique is usually called simple, or bivariate, regression. When multiple independent variables are involved, it is referred to as multiple regression. The general form of the prediction equation is

$$Y_i = b_0 + b_1 * X_{i1} + b_2 * X_{i2} + \ldots + b_j * X_{ij} + E_i$$

where Y_i and the X_{ij}'s are elements of vectors of data values for the dependent and independent variables, b_0 is the intercept, the b_j's are the other coefficients or parameters of the equation to be estimated using the least squares estimation method, and E_i is a vector of errors in predicting the Y_i's. The asterisk stands for multiplication.

The degree of accuracy of the predictions is measured by Pearson's r^2 in the bivariate case and by the Multiple R^2 in the multiple regression situation. While there are several PROCs in SAS that can be employed to do regression analyses, the one used most often is PROC REG.

Visualizing Bivariate Relationships Using PROC PLOT

PROC PLOT is useful for plotting data points for two continuous variables to check linearity assumptions and identify outliers. Although it is possible to produce more elaborate plots with PROC GPLOT, the basic syntax for plotting points is identical to that for PROC PLOT.

The basic syntax for PROC PLOT is illustrated in Exhibit 9.11. The input SAS dataset contains some fictitious data generated just for this example. There may be more than one PLOT statement per PROC PLOT statement. Multiple plot requests may be specified on each PLOT statement. To request a plot for a particular pair of variables, specify the SAS name of the Y or vertical axis variable, follow it with an asterisk (*), and then give the name of the X or horizontal axis variable. The variables in this dataset were named `x`, `y`, and `y2`.

Exhibit 9.11

Output from PROC PLOT

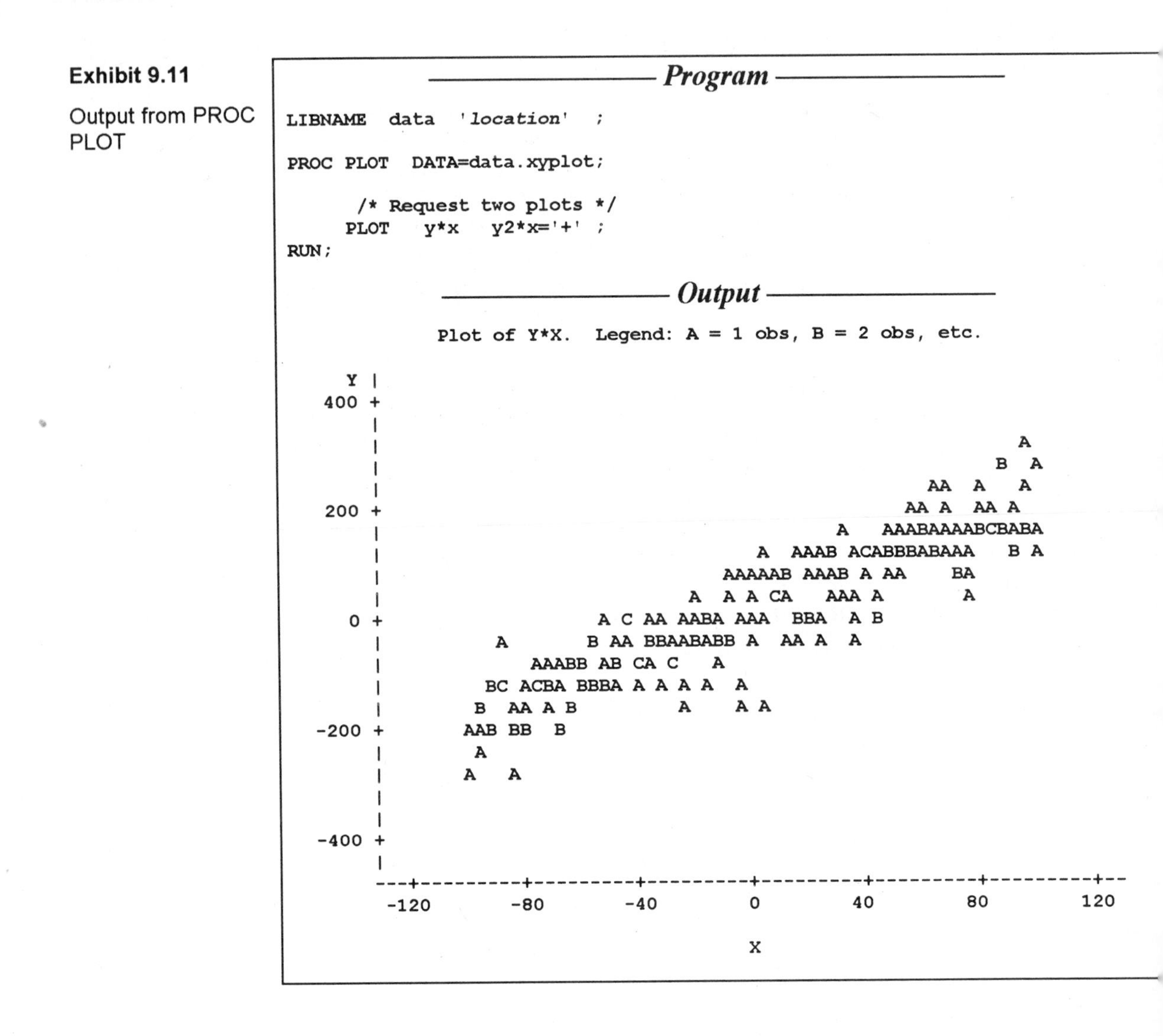

Program

```
LIBNAME  data  'location'  ;

PROC PLOT  DATA=data.xyplot;

      /* Request two plots */
     PLOT    y*x   y2*x='+' ;
RUN;
```

Output

```
            Plot of Y*X.  Legend: A = 1 obs, B = 2 obs, etc.

    Y |
  400 +
      |
      |                                                              A
      |                                                            B  A
      |                                                      AA  A   A
  200 +                                                    AA A  AA A
      |                                              A   AAABAAAABCBABA
      |                                       A  AAAB ACABBBABAAA   B A
      |                                    AAAAAB AAAB A AA    BA
      |                                A  A A CA   AAA A       A
    0 +                        A C AA AABA AAA  BBA  A B
      |               A       B AA BBAABABB A  AA A  A
      |                  AAABB AB CA C   A
      |              BC ACBA BBBA A A A A  A
      |             B  AA A B         A    A A
 -200 +            AAB BB  B
      |             A
      |            A   A
      |
      |
 -400 +
      |
      ---+---------+---------+---------+---------+---------+---------+--
       -120      -80       -40        0        40        80       120

                                      X
```

Exhibit 9.11

Output from PROC PLOT (continued)

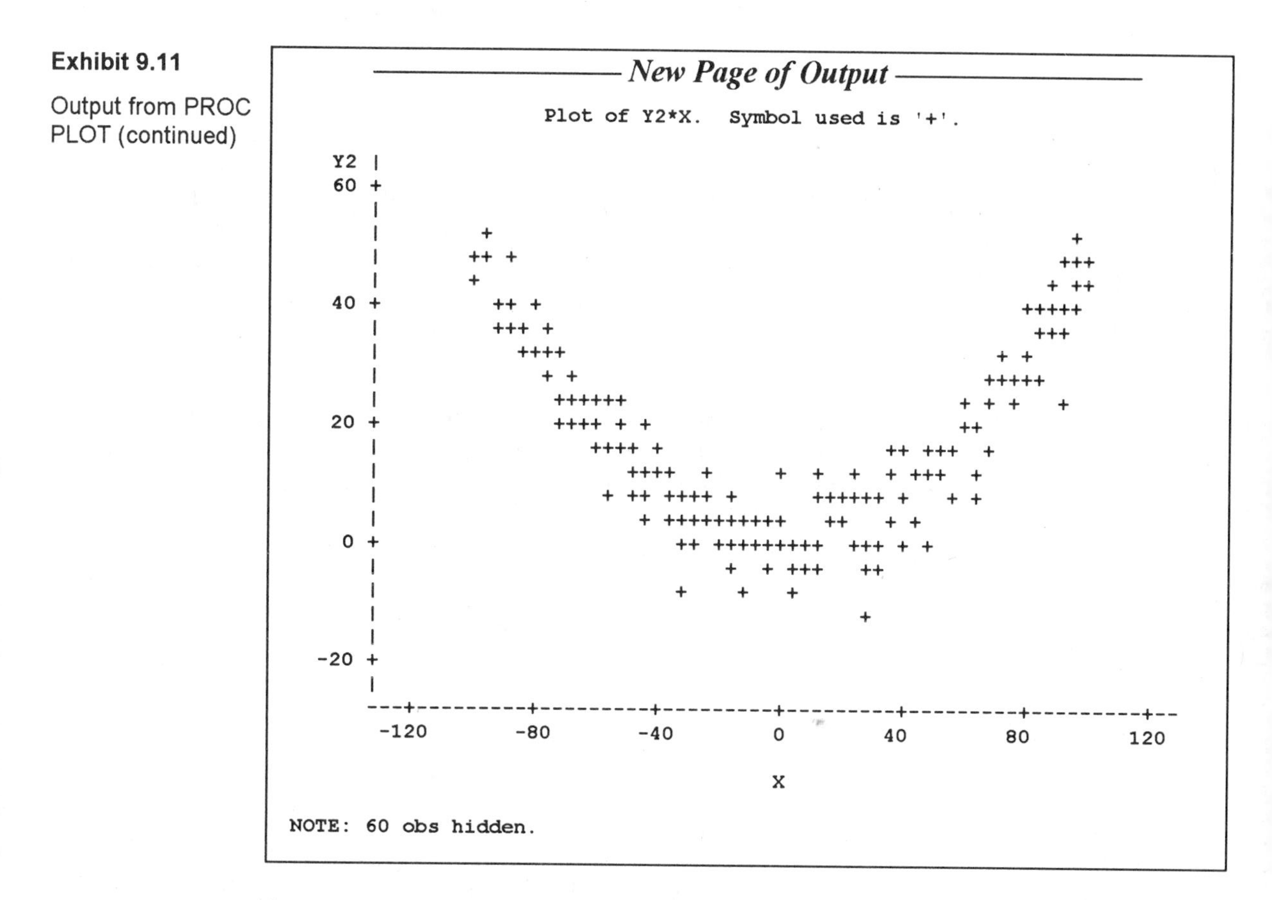

As you can see from the output for the first plot specification (`y*x`) shown in Exhibit 9.11, PROC PLOT uses letters to plot the x, y value pairs. An `A` represents one observation at a particular x-y coordinate, a `B`, two observations, a `C`, three observations, and so on.

If you feel that PROC PLOT's use of letters produces a plot that's a bit cluttered and prefer to use a particular character, such as the + sign, instead of letting SAS use letters, you may add an equal sign to the plot request, followed by your chosen character enclosed in quotes. This syntax is illustrated in the second plot request in the example above. Because PROC PLOT can no longer inform you about the number of observations at a particular point when you use a single character for plotting points, it issues the message `NOTE: 60 obs hidden.` in the output for the second plot. This is a warning that some points plotted with the + sign represent more than one observation.

Examination of a plot may reveal a clear nonlinear relationship like that in the second plot in Exhibit 9.11. It is sometimes possible to use a mathematical transformation of a variable in the model to produce a more clearly linear plot. The values of the transformed variable may also be used in the estimation of the regression equation instead of, or in addition to, the original values.

Some common transforms are to square a variable or to take its natural log or square root. These transformations can be performed by creating new SAS variables in a DATA step preceding the PROC PLOT. These new variable names may then be used on the PLOT statement to see if a more linear pattern emerges and, if necessary, on a MODEL statement in subsequent executions of PROC REG. You are referred to Daniel and Wood (1980), in the references for this chapter, for further details on the use of various linearizing transformations for regression.

If the plot reveals outliers, check the original data to make sure they were not created by a data entry error and consider whether or not they should be excluded from the analysis. Exclusion can be done with a WHERE statement following a PROC PLOT or PROC REG statement (WHERE was discussed in Chapter 6). Remember that excluding apparent outliers from the data prior to estimating the regression equation is not always appropriate. It is often better to use more sophisticated outlier identification techniques based on residual and other plots that will be covered later in this section.

Estimating Regression Models with PROC REG

PROC REG may be used to obtain least squares estimates of the coefficients for a simple or multiple regression model, as well as a Pearson r^2 or Multiple R^2 measure of the fit of the model to the data and various significance tests for the degree of fit and the coefficients. Consider the example in Exhibit 9.12. It estimates a simple regression model for predicting first-year grade point average in a postgraduate degree program. The independent variables are verbal and quantitative scores on the Graduate Record Examinations. The data are adapted from Pedhazur (1982), page 138.

Regression Syntax The MODEL statement is used to specify the dependent and independent variables to be used in the regression model to be estimated. Multiple MODEL statements may be used to estimate two or more models within a single invocation of PROC REG. On any MODEL statement, the SAS variable named to the left of the equal sign is the dependent variable. Any variable names listed to the right of the equal sign will be used as independent variables.

The STB option on the MODEL statement asks PROC REG to produce the standardized parameter estimates in addition to the usual or unstandardized parameter estimates. Social scientists often use these standardized estimates, sometimes referred to as beta or path coefficients, to compare effects of variables corrected for their scales measurement in what is known as a *path model*.

Exhibit 9.12

Estimating a Simple Regression Model with PROC REG

Program

```
LIBNAME  data  'location' ;

DATA  data.gradgpa;

      INPUT  student  gpa  grev  greq  rating $ ;
      LABEL  student = 'Student ID'
             gpa     = 'First Year GPA'
             grev    = 'Verbal GRE Score'
             greq    = 'Quantitative GRE Score'
             rating  = 'Faculty Rating' ;
DATALINES;
 1    3.2     540     625      Low
 2    4.1     680     575      Hi
 3    3.0     480     520      Low
 4    2.6     520     545      Mid
 5    3.7     490     520      Mid
 6    4.0     535     655      Mid
 7    4.3     720     630      Hi
 8    2.7     500     500      Mid
 9    3.6     575     605      Hi
10    4.1     690     555      Mid
11    2.7     545     505      Mid
12    2.9     515     540      Low
13    2.5     520     520      Mid
14    3.0     710     585      Low
15    3.3     610     600      Hi
16    3.2     540     625      Low
17    4.1     680     575      Hi
18    3.0     480     520      Low
19    2.6     520     545      Mid
20    3.7     490     520      Mid
21    4.0     535     655      Mid
22    4.3     720     630      Hi
23    2.7     500     500      Mid
24    3.6     575     605      Hi
25    4.1     690     555      Mid
26    2.7     545     505      Mid
27    2.9     515     540      Low
28    2.5     520     520      Mid
29    3.0     710     585      Low
30    3.3     610     600      Hi
PROC REG  ;
     MODEL  gpa = grev greq  / STB ;

TITLE1 'Predicting First Year Graduate GPA from' ;
TITLE2 'Verbal and Quantitative GRE Scores'      ;
RUN;
```

Exhibit 9.12

Estimating a Simple Regression Model with PROC REG (continued)

Output

```
Predicting First Year Graduate GPA from
Verbal and Quantitative GRE Scores

Model: MODEL1
Dependent Variable: GPA          First Year GPA


ANOVA

                        Sum of          Mean
Source          DF     Squares        Square       F Value       Prob>F

Model            2     5.06333       2.53166        12.726       0.0001
Error           27     5.37134       0.19894
C Total         29    10.43467

     Root MSE         0.44603     R-square          0.4852
     Dep Mean         3.31333     Adj R-sq          0.4471
     C.V.            13.46153


Parameter Estimates

                    Parameter        Standard     T for H0:
Variable  DF         Estimate           Error   Parameter=0     Prob > |T|

INTERCEP   1        -1.286698      0.97652072        -1.318         0.1987
GREV       1         0.002733      0.00112876         2.421         0.0225
GREQ       1         0.005356      0.00192781         2.778         0.0098

                 Standardized  Variable
Variable  DF         Estimate    Label

INTERCEP   1       0.00000000  Intercept
GREV       1       0.37826900  Verbal GRE Score
GREQ       1       0.43409398  Quantitative GRE Score
```

Interpreting Regression Output The basic output from PROC REG is presented in Exhibit 9.12. It is divided into three parts, with headings of `Model:`, `ANOVA`, and `Parameter Estimates`. In the upper left-hand corner under the label `Model: MODEL1`, PROC REG gives the name of the dependent variable specified on the MODEL statement. If you use multiple MODEL statements and don't supply model labels, each one will generate a separate regression output with `MODEL1` for the first label, `MODEL2` for the second, and so on.

The `ANOVA` section provides information on the fit of the estimated regression model to the data. The first table in this section provides the degrees of freedom under the heading `DF`, corrected sums of squares under the heading `Sums of Squares`, and mean squares under the heading `Mean Square`, for the sum of squares explained by the estimated regression equation and the residual, or error, sum of squares.

The explained sum of squares is given in the row labeled `Model`, under the column headed `Source`, and the same information for the residual sum of squares left unexplained by the model is presented in the row labeled `Error`, in the `Source` column. The row titled `C Total` gives the total sum of squares. This is the sum of the explained and unexplained sums of squares. The `F Value` and `Prob>F` columns give the familiar F statistic and associated level of significance for the null hypothesis that the Multiple R equals zero. This null hypothesis may also be phrased as follows: "All the parameters of the model except the intercept equal zero."

You should be aware that PROC REG uses a "listwise" deletion method for dealing with missing values. This method uses only observations with complete data on all variables to estimate a regression equation and provide associated statistics. If multiple MODEL statements are used within a PROC REG, only observations with complete data on *all* variables specified in *all* MODEL statements are included in the estimates for regression equations specified in *any* MODEL statement. The total degrees of freedom for each regression equation should be checked carefully to make sure that this method of handling missing values has not had an adverse impact on the analysis.

The second table in the ANOVA section supplies the value of the multiple R^2, which is labeled `R-square`. This figure is the proportion of the total sum of squares explained by the regression model. It is used to assess the goodness of fit of the model to the data. To obtain the value of the multiple R, simply take the positive square root of the multiple R^2. For simple bivariate regression models, this figure is equivalent to the Pearson r^2, and its positive square root is the absolute value of r. The sign of r for bivariate regressions is the same as the sign of the slope estimate for the independent variable found in the `Parameter Estimates` section of the output.

The `Parameter Estimates` section provides information about the least squares estimates of the regression equation coefficients. It also contains significance tests for the null hypothesis that each coefficient is not different from zero. Each independent variable named on a MODEL statement is listed under the column heading `Variable`. PROC REG inserts the term `INTERCEP` as its name for the intercept, or b_0, parameter. Associated with each variable name is the least squares estimate of the corresponding regression coeffient under the heading `Parameter Estimate`, the standard error of the estimated coefficient under the heading `Standard Error`, the value of the t statistic for the test of equality to zero under the heading `T for H0: Parameter=0`, and the one-tailed probability values associated with the value of the t statistic under the heading `Prob > |T|`. The final column provides the standardized estimates of the equation coefficients requested by using the STB option.

The estimated regression equation can easily be constructed from the `Parameter Estimates` column. In Exhibit 9.12, it would be

$$GPA_i = -1.2867 + .0027*GREV_i + .0054*GREQ_i + E_i,$$

rounded to four decimal places. The standardized regression equation is easily obtained from the column labeled `Standardized Estimate`. The standard errors of the coefficients can be used to calculate confidence intervals around their corresponding estimates using the standard formula

$$\text{Confidence Interval} = \text{Estimate} \pm (\text{t value} * \text{Standard Error}).$$

In this case, a 95% interval for the estimate for GREV would be

$$.0027 \pm (2.055 * .0011),$$

where 2.055 is the value of t for the two-tailed .05 level of significance for 27 degrees of freedom.

Checking Fit and Assumptions Using the PLOT Statement After estimating a regression equation, it is usually advisable to examine plots of the errors in prediction, usually called *residuals*. The most commonly used diagnostic plot, which plots the residuals or studentized residuals against the predicted values, may be used to check for outliers, assess the equality of variances assumption, and check for possible curvilinear relationships or omitted variables. If such plots reveal a random scattering of points with no clearly outlying points, the estimated model may be adequate to characterize the data. However, if outliers are found or if the points exhibit a curvilinear, funnel-shaped, oscillating, or other reasonably clear pattern, you should refine the initial model.

Plots of residuals against the independent variables of a multiple regression model are often used to reveal which variable or variables have a curvilinear relationship with the dependent variable or are linearly correlated with important omitted variables. Plots of more sophisticated diagnostic statistics, detailed in Belsey, Kuh, and Welsch's *Regression Diagnostics,* can also be employed to further refine a regression analysis. It is not possible to discuss the derivation and proper interpretation of each of these diagnostic statistics in this book. You should consult Belsey, Kuh, and Welsch (1980), Freund and Littell (1991), Chapter 3, or Meyers (1990) for thorough treatments of them.

It is quite easy to construct diagnostic plots within PROC REG using the PLOT statement. This statement must follow a MODEL statement, and produces output that applies only to the regression equation estimated in that model. PROC REG automatically generates the values of residuals, predicted values, and other diagnostic statistics for plotting. Each statistic has a corresponding SAS variable name, ending in a period, which may be used on a PLOT statement. For example, SAS gives the name

PREDICTED. (which may also be abbreviated as P. or PRED.) to the predicted values for the regression equation. It assigns the name RESIDUAL. (which may be abbreviated as R.) to the residuals. Exhibit 9.13 shows the SAS variable names given to some commonly used diagnostic statistics. The names may be either in uppercase or in lowercase. It is very important to remember the period.

The syntax of the PLOT statement in PROC REG is very similar to that of the PLOT statement used in PROC PLOT. To construct a single plot of the residuals versus the predicted values, simply use their names or abbreviated names on a PLOT statement as follows:

```
PLOT RESIDUAL. * PREDICTED. ;
```

or

```
PLOT R. * P. ;
```

The PLOT statement below requests two separate plots. The HPLOTS=2 option, which sets the number of horizontal plots, and the VPLOTS=1 option, which sets the number of vertical plots, used together tell SAS to place the two plots side by side on a single page. Using these options together can facilitate comparison of plots and can also save paper.

```
PLOT (RESIDUAL. STUDENT.) * PRED.
     /  HPLOTS=2  VPLOTS=1 ;
```

The values of each of the diagnostic statistics named within the parentheses will be plotted on the Y axis with the predicted values plotted on the X axis.

Exhibit 9.13

Diagnostic Plotting Variables for PROC REG

Full Name	Abbreviation	Definition
PREDICTED.	PRED. or P.	Predicted value of Y
RESIDUAL.	R.	Residual (Y - Yhat)
STUDENT.	STUDENT.	Standardized Residual (Residual/ Std. Err. Residual)

Checking for Correlated Errors Using the DW Option A key assumption in regression analysis is that the errors are independent of each other and therefore exhibit no autocorrelation. This assumption almost always needs to be checked when the data are measurements of the same variables on a single entity over time — the level of a variable at any given time is usually fairly close to its level in preceding time periods. A regression analysis of national economic data, such as quarterly consumer expenditures as a function of average wages, GDP, and money supply, would be an example of this type of regression analysis.

The Durban-Watson D statistic is often used to test for the presence of first-order autocorrelation in regression models involving time series data. Its derivation and use are covered in almost every introductory econometrics text. It can be obtained from PROC REG by adding the DW option to the MODEL statement. If the Durban-Watson test indicates that autocorrelation is present in the data, it may be best to use PROC AUTOREG rather than PROC REG to construct an appropriate regression model.

Checking for Multicollinearity with the VIF and COLLINOINT Options
When multiple regression models have many independent variables, it is not unusual for some or all of them to be highly correlated. This condition is referrred to as multicollinearity. The model's parameter estimates have high standard errors and, therefore, may have incorrect signs and inappropriate magnitudes or lack statistical significance. (Meyers, 1990, Section 3.7.) If a researcher is interested in interpreting the parameter estimates as measures of the structural or causal relationships between the dependent and independent variables, poor estimates caused by multicollinearity can result in misleading conclusions about the underlying causal processes.

PROC REG provides two useful options on the MODEL statement for diagnosing the presence and pattern of multicollinearity: VIF and COLLINOINT. The VIF option will produce variance inflation factors for each variable. It measures how much multicollinearity in the data has increased the variance of the parameter estimate for that variable. The COLLINOINT option provides the eigenvalues and corresponding eigenvectors of the correlation matrix of the independent variables. This can help identify sets of variables that are highly intercorrelated. Interpreting the output for these options and dealing with multicollinearity in regression models are beyond the scope of this book. Useful discussions can be found in many texts on regression analysis, including Belsey, Kuh, and Welsch (1980), Freund and Littell (1991), Chapter 4, Kleinbaum, Kupper, and Muller (1988), Section 12-5, and Meyers (1990), Chapter 8.

Using PROC REG Interactively Some SAS data analysis PROCs may be run interactively. This means that if you're using SAS in interactive mode (see Appendix A), you need not submit all the statements for a PROC for execution at one time. Instead, you can submit one or more statements, examine the output produced by them, and then write and submit additional statements.

PROC REG is an interactive procedure. When using PROC REG interactively, you must remember to list all the variables that might be used in a particular regression analysis on a VAR statement. It must immediately follow the PROC REG statement. Exhibit 9.14 illustrates this concept. You should consult the documentation for using PROC REG interactively for a complete discussion of all interactive commands.

Exhibit 9.14

Syntax for Using PROC REG Interactively

```
PROC REG DATA=whatever ;         /* Execute PROC REG */
     VAR  y x1 x2 x3 x4 x5 ;     /* List all variables that might be used */
     MODEL y = x1 x2       ;     /* Estimate initial model                */
RUN;                             /* Issue SUBMIT command in PGM window    */

                              /* Examine output for this model in
                                 OUTPUT window                            */
PLOT r.*p.   student.*p.      /* Plot resid's against predicted vals.     */
     r.*x1   r.*x2            /* Plot residuals against ind. vars.        */
   / HPLOTS=2 VPLOTS=1;       /* Arrange plots two per page               */
RUN;                          /* Issue SUBMIT command and view output     */

MODEL  y = x1 x2 x3 x4 ;      /* Add x3 and x4 to the model               */
RUN;                          /* Issue SUBMIT command and view output     */

PLOT r.*p. student.*p.        /* Plot residuals for last model            */
   / HPLOTS=2 VPLOTS=1;
RUN;                          /* Issue SUBMIT command and view output     */

QUIT;                         /* Halt interactive execution of PROC REG   */
```

Analysis of Covariance: Statistics for Both Categorical and Continuous Independent Variables

An analysis of covariance (ANCOVA) may be used when the dependent variable is continuous and the independent variables are a mixture of categorical and continuous scales. In the simplest case, researchers are usually interested in examining differences between group means of the dependent variable defined by the categorical variable or variables when the effects of one or more continuous variables are controlled. For example, we might be interested in comparing the effects of two instructional strategies with a control with respect to mastery of an academic subject, controlling for the IQ levels of the students assigned to each group.

Statisticians usually recommend that testing for differences in this manner requires testing the assumption that the relationship between the covariates and the dependent variable is the same within each group defined by the categorical variable. We will show you how to use PROC GLM to handle this basic situation. Analyses of covariance analysis strategy and interpretation can get more complex than this when the number of independent variables increases. You can learn more about handling involved analysis of covariance designs in Maxwell and Delaney (1990), Chapter 9, and Littell, Freund, and Spector (1991), Chapter 6.

Using PROC GLM for ANCOVA

The example in Exhibit 9.15 shows how to use PROC GLM to test for differences in the within-group regression coefficients, or slopes, of two continuous variables. It also conducts a separate test for differences in means between groups, adjusting for the two continuous variables. We use the data from Pedhazur (1982) shown above and use faculty rating of student prior to admission (`rating`) as our categorical variable. Verbal GRE score (`grev`) and quantitative GRE score (`greq`) are the covariates. The question of interest might be whether or not faculty ratings can predict graduate school performance (`gpa`) after verbal and quantitative GRE scores are taken into account.

Two PROC GLM statements are used here because only one MODEL statement is allowed per PROC GLM statement and we must estimate two models for this analysis. Since `rating` is a categorical variable, its name appears on the CLASS statement for both PROC GLMs. The "main effects" of `greq`, `grev`, and `rating` appear on both MODEL statements. We also include two interaction terms on the first statement to test the equality of within-group slopes: `rating*greq` and `rating*grev`. These are the terms added to the basic ANCOVA model specified in the second MODEL statement for the differences between the average slopes for `greq` and `grev` and the within-group slopes for these variables. The SOLUTION option on the second MODEL statement tells PROC GLM to print out the coefficients of the linear model. These coefficents will include the estimates of the average slopes for `greq` and `grev`.

Exhibit 9.15

Performing ANCOVA with PROC GLM

Program

```
LIBNAME  data  'location' ;

/* Test for equality of slopes */
PROC GLM  DATA=data.gradgpa ;

  /* Faculty rating is the categorical variable */
  CLASS  rating ;

  /* Interaction terms test equality of slopes */
  MODEL  gpa  = greq grev   rating
                rating*greq rating*grev  ;

TITLE1  'Test of Homogeneity of Slopes for' ;
TITLE2  'ANCOVA with Two Covariates' ;
RUN;

/* Test for equality of means adjusted for covariates */
PROC GLM  DATA=data.gradgpa ;
  CLASS  rating ;
  /* Test for the common slopes model */
  MODEL  gpa  =  greq  grev  rating / SOLUTION ;

TITLE1  'Test of Equality of Group Means' ;
TITLE2  'Assuming Equal Slopes for Two Covariates' ;
RUN;
```

Exhibit 9.15

Performing ANCOVA with PROC GLM (continued)

Output

```
Test of Homogeneity of Slopes for
ANCOVA with Two Covariates

General Linear Models Procedure
Class Level Information

Class      Levels    Values

RATING          3    Hi Low Mid


Number of observations in data set = 30
```

New Page of Output

```
Test of Homogeneity of Slopes for
ANCOVA with Two Covariates

General Linear Models Procedure

Dependent Variable: GPA    First Year GPA

                                   Sum of            Mean
Source                 DF         Squares          Square  F Value      Pr > F

Model                   8       7.4502955       0.9312869     6.55      0.0003
Error                  21       2.9843712       0.1421129
Corrected Total        29      10.4346667

                 R-Square            C.V.        Root MSE               GPA Mean
                 0.713995        11.37763          0.3770                 3.3133

Source                 DF       Type I SS     Mean Square  F Value      Pr > F

GREQ                    1       3.8973617       3.8973617    27.42      0.0001
GREV                    1       1.1659642       1.1659642     8.20      0.0093
RATING                  2       0.7782440       0.3891220     2.74      0.0878
GREQ*RATING             2       0.6618389       0.3309194     2.33      0.1221
GREV*RATING             2       0.9468867       0.4734434     3.33      0.0554

Source                 DF     Type III SS     Mean Square  F Value      Pr > F

GREQ                    1       0.2163008       0.2163008     1.52      0.2309
GREV                    1       1.2215347       1.2215347     8.60      0.0080
RATING                  2       0.6919992       0.3459996     2.43      0.1120
GREQ*RATING             2       0.2786862       0.1393431     0.98      0.3916
GREV*RATING             2       0.9468867       0.4734434     3.33      0.0554
```

Exhibit 9.15

Performing ANCOVA with PROC GLM (continued)

New Page of Output

```
Test of Equality of Group Means
Assuming Equal Slopes for Two Covariates

General Linear Models Procedure
Class Level Information

Class     Levels    Values

RATING         3    Hi Low Mid

                 Number of observations in data set = 30
```

New Page of Output

```
Test of Equality of Group Means
Assuming Equal Slopes for Two Covariates

General Linear Models Procedure

Dependent Variable: GPA   First Year GPA
                                   Sum of          Mean
Source                  DF        Squares        Square  F Value      Pr > F

Model                    4      5.8415699     1.4603925     7.95      0.0003
Error                   25      4.5930968     0.1837239
Corrected Total         29     10.4346667

               R-Square             C.V.       Root MSE             GPA Mean

               0.559823         12.93653         0.4286               3.3133


Source                  DF      Type I SS   Mean Square  F Value      Pr > F

GREQ                     1      3.8973617     3.8973617    21.21      0.0001
GREV                     1      1.1659642     1.1659642     6.35      0.0185
RATING                   2      0.7782440     0.3891220     2.12      0.1413

Source                  DF    Type III SS   Mean Square  F Value      Pr > F

GREQ                     1      1.3840019     1.3840019     7.53      0.0111
GREV                     1      0.6983293     0.6983293     3.80      0.0625
RATING                   2      0.7782440     0.3891220     2.12      0.1413
```

Exhibit 9.15

Performing ANCOVA with PROC GLM (continued)

```
                     Parameter          Std Error of      T for H0:
                     Estimate           Estimate          Parameter=0       Pr > |T|

INTERCEPT           -1.025067195 B      1.09648134           -0.93          0.3588
GREQ                 0.005458182        0.00198867            2.74          0.0111
GREV                 0.002298521        0.00117897            1.95          0.0625
RATING      Hi       0.076093517 B      0.23862121            0.32          0.7525
            Low     -0.337495848 B      0.19587473           -1.72          0.0972
            Mid      0.000000000 B       .                     .              .

NOTE: The X'X matrix has been found to be singular and a generalized
      inverse was used to solve the normal equations.   Estimates
      followed by the letter 'B' are biased, and are not unique
      estimators of the parameters.
```

Interpreting PROC GLM ANCOVA Output

Since you are already familiar with the basic layout for PROC GLM output, we will only highlight those aspects of the output for this example that are particularly relevant to the analysis of covariance. On the second page of the output for the first PROC GLM, you will notice that the Multiple R^2 is a respectable .71 and is significantly different from zero at the .0003 level of significance. The hypothesis tests based on Type III sums of squares are independent of the ordering of the variables on the MODEL statement. They provide useful tests for the two terms in the model that bear on the equality of slopes. You can see that when the effects of all other terms in the model are controlled, neither of the within-group slope terms achieves significance at the .05 level, although the `grev*rating` term does approach this level. Thus, it is possible to conclude that the within-group slopes are equal and that the simpler average-slopes ANCOVA model is appropriate for this data.

Moving on to the fourth page, you can see that the simple ANCOVA model explains a bit less variance. The R^2 is .56, but is still significantly different from zero ($P \leq .0003$). The Type III sum of squares for rating, however, is not significant, indicating that when the effects of the GRE scores are controlled, faculty ratings do not help predict mean differences in performance in graduate school. If the test for this term had achieved significance, you would have been led to the opposite conclusion.

The parameter estimates of the linear model specified on the MODEL statement follow the Type III sums of squares table. Note that the coefficients estimated for `greq` and `grev` are not marked as biased and are appropriate estimates of the average slopes for these two variables. The reason the estimates for the intercept and the categories of `rating` are biased is quite technical and is beyond the scope of this book. (If you would like to explore this issue further, see Littell, Freund, and Spector (1991), Chapter 4.) The fact that they are marked as biased, however, does not invalidate the calculation of any of the sums of squares or significance tests presented in the preceding tables nor the estimates for the slopes of the covariates.

Using PROC REG for ANCOVA

The PROC REG approach to the analysis of covariance requires you to construct all the necessary dummy variables needed in the linear models before using PROC REG for the analysis. This approach can involve more effort than using PROC GLM. However, in more complex situations it can give you more control over model terms and tests. The program in Exhibit 9.16 shows how to approach the analysis we performed above using PROC REG.

The DATA step creates the dummy variables `hirate` and `lowrate` for comparing the Hi- and Low-rated students to the Mid-rated students. Notice that ELSE IF statements for the zero values are used to explicitly specify each value of rating that is to be coded zero. This is better practice than merely using a statement like

```
ELSE hirate=0 ;
```

because it avoids the problem of possibly including observations with missing values in the comparison group. The interaction terms for assessing the difference of the within-group slopes from the average slopes are simply created as new SAS variables that are products of the appropriate dummy variable and the two covariates.

The MODEL statement for the first PROC REG estimates the unequal slopes linear model. The TEST statement following the first MODEL statement tests the null hypothesis that all of the within-group slope interaction terms equal zero. If this hypothesis is not rejected, the average slope model specified in the second MODEL statement can be considered. Note that the average slope model contains the two dummy variables `hirate` and `lowrate`, but none of the product variables. The second TEST statement applies to the equal slopes model. It tests the null hypothesis that the coefficients of the two rating dummy variables equal zero, which is equivalent to a null hypothesis of no effect of faculty rating.

The use of the RUN statements allows us to employ two different TITLE2 statements to appropriately label the output for the two MODEL statements. Note that the text on the second TITLE2 statement is not the same as that on the first.

The output for this example is also presented in Exhibit 9.16. The first page shows the output for the first MODEL statement. Since its structure is similar to that of the output reviewed in detail when we discussed PROC REG above, we will not explain it in detail. The second page gives the numerator and denominator mean squares, the value of the F statistics, and the probability level for the F statistic for the test of the null hypothesis of equal within-group slopes specified on the first TEST statement. The third page displays the estimates for the coefficients of the equal slopes model and related statistics. You might wish to compare these to the corresponding estimates produced by PROC GLM, above. The results of the test for no faculty rating effect requested by the second TEST statement are shown on the fourth page. You may wish to compare them with the test for the effect of the SAS variable rating in the GLM example output.

Exhibit 9.16

Performing ANCOVA with PROC REG

Program

```
LIBNAME  data  'location' ;

/* Create dummy variables and interaction variables */
DATA  dummies ;
      SET  data.gradgpa ;

      /* Main effect dummy variable for Hi rating  vs. Mid rating */
      IF  rating = 'Hi'   THEN  hirate = 1 ;
      ELSE IF  rating = 'Low'  OR  rating = 'Mid'  THEN  hirate = 0 ;

      /* Main effect dummy variable for Low rating vs. Mid rating */
      IF  rating = 'Low'  THEN  lowrate= 1 ;
      ELSE IF  rating = 'Hi'   OR  rating = 'Mid'  THEN lowrate = 0 ;

      /* Within-group slope interaction terms */
      greqhi = hirate*greq ;   greqlow = lowrate*greq ;
      grevhi = hirate*grev ;   grevlow = lowrate*grev ;


      LABEL   hirate = 'Hi vs. Mid faculty rating'
             lowrate = 'Low vs. Mid faculty rating'
              greqhi = 'Hi rating - quantitative GRE'
             greqlow = 'Low rating - quantitative GRE'
              grevhi = 'Hi rating - verbal GRE'
             grevlow = 'Low rating - verbal GRE'  ;
RUN;

PROC REG  DATA = dummies ;

/* Compare R squared for average slopes model to within-group slopes
   model. Always place full MODEL statement before restricted MODEL
   statement. */

MODEL gpa = greq  grev  hirate  lowrate    /* Full model with within */
          greqhi greqlow  grevhi grevlow; /* group slope interactions */

/* Test that all within-group slopes are equal to the average slopes */
TEST greqhi = greqlow = grevhi = grevlow = 0 ;

TITLE1 'ANCOVA with Two Covariates using PROC REG';
TITLE2 'Different Within-Group Slopes Model' ;
RUN;

MODEL gpa = greq  grev  hirate  lowrate;     /* Average slopes model */
/* Test for faculty rating adjusted for GRE score */
TEST hirate=lowrate=0 ;

TITLE2 'Average Slopes Model' ;
RUN;
```

Exhibit 9.16

Performing ANCOVA with PROC REG (continued)

Output

```
ANCOVA with Two Covariates using PROC REG
Different within Group Slopes Model

Model: MODEL1
Dependent Variable: GPA        First Year GPA

ANOVA

                          Sum of           Mean
Source          DF       Squares          Square       F Value       Prob>F

Model            8       7.45030         0.93129         6.553       0.0003
Error           21       2.98437         0.14211
C Total         29      10.43467

    Root MSE         0.37698      R-square          0.7140
    Dep Mean         3.31333      Adj R-sq          0.6050
    C.V.            11.37763

Parameter Estimates

                    Parameter        Standard     T for H0:
Variable   DF        Estimate           Error    Parameter=0    Prob > |T|

INTERCEP    1       -3.149424      1.29659777         -2.429        0.0242
GREQ        1        0.007370      0.00208442          3.536        0.0020
GREV        1        0.004300      0.00164414          2.615        0.0162
HIRATE      1        2.998823      4.35469140          0.689        0.4986
LOWRATE     1        4.955616      2.27006120          2.183        0.0405
GREQHI      1       -0.007334      0.00731344         -1.003        0.3274
GREQLOW     1       -0.004682      0.00422209         -1.109        0.2800
GREVHI      1        0.001818      0.00291235          0.624        0.5392
GREVLOW     1       -0.004847      0.00235689         -2.056        0.0524

                 Variable
Variable   DF      Label

INTERCEP    1   Intercept
GREQ        1   Quantitative GRE Score
GREV        1   Verbal GRE Score
HIRATE      1   Hi vs. Mid faculty rating
LOWRATE     1   Low vs. Mid faculty rating
GREQHI      1   Hi rating - quantitative GRE
GREQLOW     1   Low rating - quantitative GRE
GREVHI      1   Hi rating - verbal GRE
GREVLOW     1   Low rating - verbal GRE
```

New Page of Output

```
ANCOVA with Two Covariates using PROC REG
Different within Group Slopes Model

Dependent Variable: GPA
Test:          Numerator:       0.4022  DF:    4   F value:    2.8300
               Denominator:   0.142113  DF:   21   Prob>F:     0.0506
```

Exhibit 9.16

Performing ANCOVA with PROC REG (continued)

New Page of Output

```
ANCOVA with Two Covariates using PROC REG
Average Slopes Model

Model: MODEL2
Dependent Variable: GPA        First Year GPA

ANOVA

                        Sum of         Mean
Source          DF     Squares       Square      F Value       Prob>F

Model            4     5.84157      1.46039        7.949       0.0003
Error           25     4.59310      0.18372
C Total         29    10.43467

    Root MSE         0.42863     R-square        0.5598
    Dep Mean         3.31333     Adj R-sq        0.4894
    C.V.            12.93653

Parameter Estimates

                      Parameter       Standard    T for H0:
Variable    DF        Estimate           Error    Parameter=0    Prob > |T|

INTERCEP     1       -1.025067      1.09648134         -0.935        0.3588
GREQ         1        0.005458      0.00198867          2.745        0.0111
GREV         1        0.002299      0.00117897          1.950        0.0625
HIRATE       1        0.076094      0.23862121          0.319        0.7525
LOWRATE      1       -0.337496      0.19587473         -1.723        0.0972

                   Variable
Variable    DF      Label

INTERCEP     1   Intercept
GREQ         1   Quantitative GRE Score
GREV         1   Verbal GRE Score
HIRATE       1   Hi vs. Mid faculty rating
LOWRATE      1   Low vs. Mid faculty rating
```

New Page of Output

```
ANCOVA with Two Covariates using PROC REG
Average Slopes Model

Dependent Variable: GPA
Test:          Numerator:      0.3891   DF:    2   F value:    2.1180
               Denominator:  0.183724   DF:   25   Prob>F:     0.1413
```

Recap

Syntax Summary

Here's a summary of the syntax of the statistical procedures needed to implement the statistical techniques you've learned about in this chapter. (See the syntax description in Chapter 1 for an explanation of the notation.)

PROC CHART for creating frequency distributions within groups:

```
PROC CHART DATA=SAS-data-set;
      HBAR | VBAR depvarname1 <depvarname2 ...>
      </ GROUP= groupvar> ;
```

PROC UNIVARIATE for creating side-by-side box plots:

```
PROC UNIVARIATE DATA=SAS-data-set <PLOT> ;
      VAR varname1 <varname2 ...> ;
      BY groupvar ;*
```

* Note: Dataset must be sorted by *groupvar*.

PROC TTEST for testing difference means for two independent groups:

```
PROC TTEST <DATA=SAS-data-set> ;
      CLASS groupvar ;
      VAR varname1 <varname2 ...> ;
```

PROC GLM for ANOVA and ANCOVA:

```
PROC GLM <DATA=SAS-data-set> ;
CLASS groupvar1 <groupvar2 ...> ;
       MODEL depvar = model-specification
       </ <SOLUTION>> ;
<MEANS effect </ <BON> <DUNCAN> <LSD>
                    <SCHEFFE> <SNK> <TUKEY>>> ;
<CONTRAST 'label' effect contrast-spec;>
```

A model-specification is a list of one or more individual variable names, or variable names separated by an * or a ||, or a nested specification of the form *nestedvar (nestingvar)*. An *effect* is one component of a model specification. See the discussion of the CONTRAST statement in this chapter for an explanation of *contrast-spec*.

PROC MEANS for testing difference between means for related groups:

```
PROC MEANS DATA=SAS-data-set
      <N> <MEAN> <STDERR> <PRT> ;
VAR varname1 <varname2 ...> ;*
```

*Note: Variable names must be SAS variables that are the differences of two SAS variables representing the original variables being compared.

PROC GLM for simple repeated measures design:

```
PROC GLM DATA=SAS-data-set ;
      MODEL var1 var2 <var3 ...> =
      </ NOUNI> ;
REPEATED factorname </ NOU SUMMARY>;
```

PROC PLOT for X-Y plots:

```
PROC PLOT DATA=SAS-data-set ;
PLOT yvar * xvar | yvar * xvar='plotchar' ;*
```

* Note: A `plotchar` may be any one of the printable characters available on your system. Also, you may place more than one plot request on a PLOT statement.

PROC REG for regression or ANCOVA:

```
PROC REG <DATA=SAS-data-set> ;
      <VAR varname1 varname2 <varname3 ...> ;>
MODEL yvarname = xvarname1 <xvarname2 ...>
       </ STB> ;
<PLOT yvar1 * xvar1 </ <HPLOTS=n> <VPLOTS=m>>;*
```

* Note: Either the `yvar1` or the `xvar1` may be PREDICTED., RESIDUAL., or STUDENT. . More than one plot request may appear on a PLOT statement. *n* and *m* are integer numbers.

Content Review

In this chapter, we have introduced you to the use of some of the SAS System's powerful PROCs for computing descriptive and inferential statistics for independent group t tests, basic ANOVA, related group t tests, repeated measures ANOVA, bivariate and multiple regression, and basic analysis of covariance. We have by no means exhaustively covered each of these topics or fully explored the capacities of such powerful procedures as PROC GLM and PROC REG. You can pursue more advanced uses of these procedures by reading the references cited below or by taking more advanced statistics courses.

The common thread linking all of the statistical techniques covered in this chapter is that the dependent variable is assumed to have continuous values. In the next chapter, we will turn our attention to SAS PROCs for situations where the dependent variable is categorical or discrete in nature.

References

Belsley, David A., Edwin Kuh, and Roy E. Welsch, *Regression Diagnostics: Identifying Influential Data and Sources of Collinearity*. New York: John Wiley and Sons, 1980.

Daniel, Cuthbert and Fred S. Wood, *Fitting Equations to Data: Computer Analysis of Multifactor Data, Second Edition*. New York: John Wiley and Sons, 1980.

Freund, Rudolf J. and Ramon C. Littell, *SAS® System for Regression, Second Edition*. Cary, NC: SAS Institute, Inc., 1991.

Freund, Rudolf J. Ramon C. Littell, and Phillip C. Spector, *SAS® System for Linear Models, Third Edition*. Cary, NC: SAS Institute, Inc., 1991.

Maxwell, Scott E. and Harold D. Delaney, *Designing Experiments and Analyzing Data: A Model Comparison Perspective*. Belmont, CA: Wadsworth Publishing Co., 1990.

Meyers, R. H., *Classical and Modern Regression with Applications, Second Edition*. Boston: PWS/Kent Publishing Co., 1990.

Statistics for Relationships with Categorical Dependent Variables

In the preceding chapter, we explained how to use the TTEST, GLM, and REG procedures to examine relationships between a continuous dependent variable and one or more continuous or categorical independent variables. This chapter is also concerned with statistics for examining relationships among variables. Here, however, we will explore techniques for dependent variables that are either categorical (i.e., measured on a nominal scale) or rank-ordered (i.e., measured on an ordinal scale). In this chapter we will

- Show you how to use PROC FREQ to obtain bivariate crosstabulations of categorical and rank-ordered variables and associated tests for and measures of relationship
- Explain the use of PROC NPAR1WAY for testing for differences between groups of observations when the dependent variable is rank-ordered
- Review the use of PROC LOGISTIC and PROC CATMOD for developing logistic regression prediction equations for dependent variables with two categories

Examining Bivariate Relationships

Producing a Crosstabulation with Tests and Measures of Association with PROC FREQ

We have already seen PROC FREQ used for creating univariate frequency distributions. It may also be used to produce counts of the combinations of values of two or more variables. either of which may be character or numeric. The printed display of the counts is called a joint frequency distribution table, a crosstabulation, or simply a crosstab.

PROC FREQ can produce a number of commonly used statistical tests and measures of association for bivariate association. When these tests and measures are requested, PROC FREQ produces them for *every* crosstabulation. It is up to you to understand the level of measurement and the sampling assumptions underlying each statistic and decide whether or not it is applicable to the crosstabulation you have created.

The code shown below illustrates the simplest way to use PROC FREQ to generate just a crosstabulation for two numeric SAS variables taken from the 1991 General Social Survey. The PROC FREQ statement used here resembles the one used to produce univariate distributions. The difference is the use of the asterisk between variable names on the TABLES statement. The variable name placed before an asterisk becomes the row variable for the crosstabulation output; the one placed after it becomes the column variable.

```
LIBNAME  DATA   'location' ;

PROC FREQ  DATA=data.gss1991 ;
     /* rowvar * colvar table request */
     TABLES  race * cappun ;
TITLE1 'Crosstab of Respondent Race and';
TITLE2 'Opinion on Death Penalty for Murder';
RUN;
```

You may request more than one crosstab on a TABLES statement by adding pairs of variables separated by asterisks. For example, specifying

```
TABLES varname1*varname2 varname1*varname3
       varname2*varname4 ;
```

would produce three separate tables. If you wish one variable crosstabulated with several others, you may use parentheses in the table specification.

For example, since the first two tables on the above statement have the same row variable name (`varname1`), it could be simplified as

```
TABLES   varname1*(varname2 varname3)
         varname2*varname4;
```

Interpreting Crosstabulation Output The first thing to notice about the output in Exhibit 10.1 is the note at the bottom left of the table. This is the number of observations in the dataset that are not included in the crosstabulation because of missing values found on one or both of the variables. Only counts and percentages for observations without missing values are displayed for the row and column variables. PROC FREQ does not include observations with a missing value on either or both of the variables in the table counts.

Also note that the body of the table contains one cell for each combination of nonmissing values of the row and column variable. Four numbers are printed within each of these cells. The key to what the numbers are is printed in the upper left-hand corner of the crosstab. Let's use this key to examine the cell for `RACE` = 1 and `CAPPUN` = 1, which the codebook for this survey tells us is the cell for white respondents favoring capital punishment.

The top number in each cell (`Frequency`) is the count of observations for that cell's combination of values. For `RACE` = 1 and `CAPPUN` = 1, this frequency is 947. Below the cell frequency is the cell's percentage of the grand total of observations (`Percent`). In our example cell, this is 947 ÷ 1507 = 62.84. The next number in each cell (`Row Pct`) is the cell's percentage of the total number of observations in that row. In the cell we're examining, this figure is 947 ÷ 1260 = 75.16. The bottom number in each cell (`Col Pct`) is the cell's percentage of the total number of observations in that column. For our example cell, we find 947 ÷1078 = 87.85 percent.

Each row and each column has a marginal total count and percentage. These numbers are the row's and column's portions of the table grand total. For example, the first-row marginal total of this table shows that 1260, or 83.61 percent, of all respondents were white. The marginal total for the first column indicates that 1078, or 71.53 percent, of respondents favored the death penalty for murder. The grand-total number of observations in the table can be found at the bottom right corner of the table, above the figure 100.00. The 100.00 figure indicates that the grand total of 1507 equals 100% of the observations included in the crosstabulation.

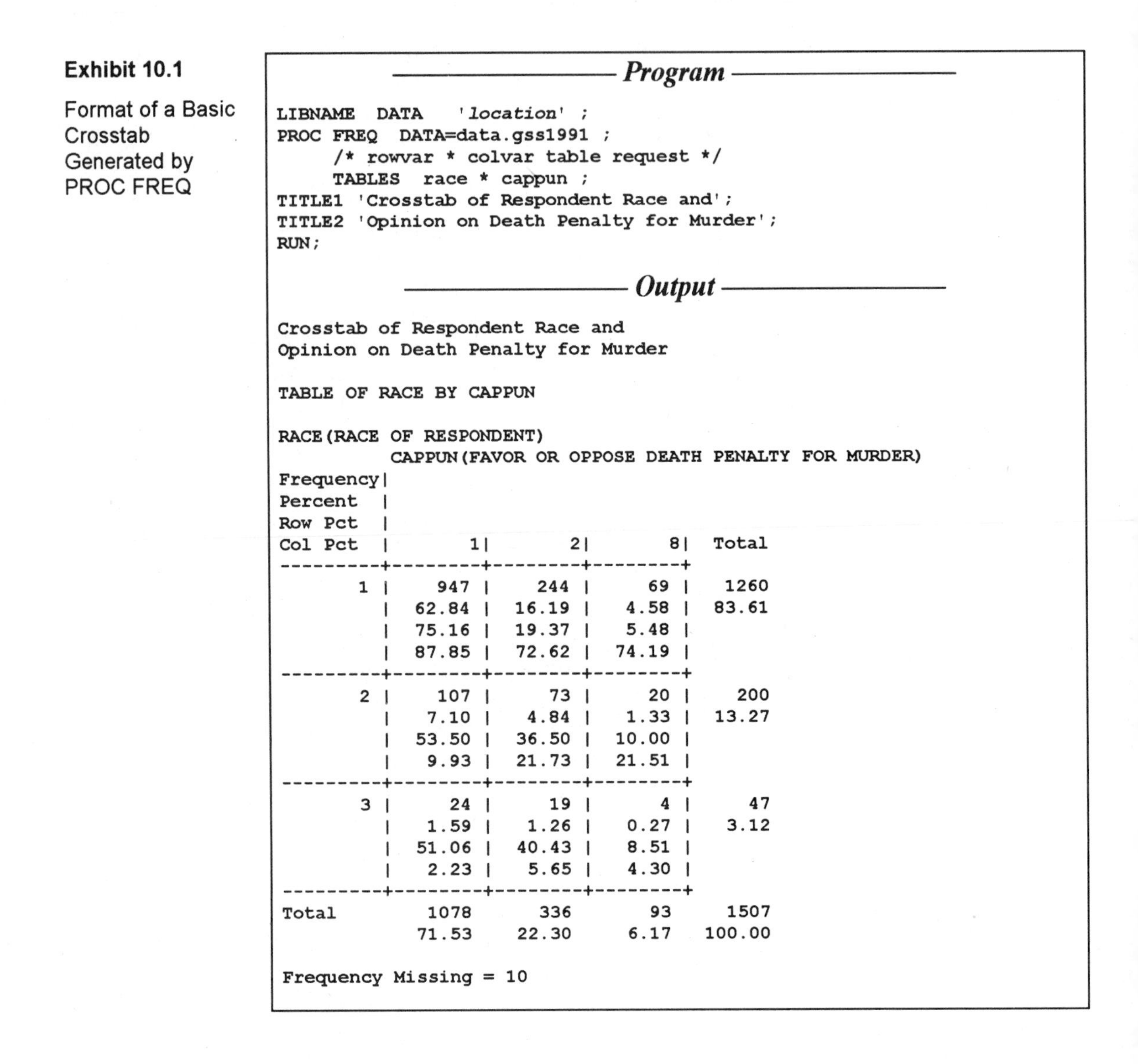

Program

```
LIBNAME  DATA   'location' ;
PROC FREQ  DATA=data.gss1991 ;
     /* rowvar * colvar table request */
     TABLES  race * cappun ;
TITLE1 'Crosstab of Respondent Race and';
TITLE2 'Opinion on Death Penalty for Murder';
RUN;
```

Output

```
Crosstab of Respondent Race and
Opinion on Death Penalty for Murder

TABLE OF RACE BY CAPPUN

RACE(RACE OF RESPONDENT)
          CAPPUN(FAVOR OR OPPOSE DEATH PENALTY FOR MURDER)
Frequency|
Percent  |
Row Pct  |
Col Pct  |       1|       2|       8|  Total
---------+--------+--------+--------+
       1 |    947 |    244 |     69 |   1260
         |  62.84 |  16.19 |   4.58 |  83.61
         |  75.16 |  19.37 |   5.48 |
         |  87.85 |  72.62 |  74.19 |
---------+--------+--------+--------+
       2 |    107 |     73 |     20 |    200
         |   7.10 |   4.84 |   1.33 |  13.27
         |  53.50 |  36.50 |  10.00 |
         |   9.93 |  21.73 |  21.51 |
---------+--------+--------+--------+
       3 |     24 |     19 |      4 |     47
         |   1.59 |   1.26 |   0.27 |   3.12
         |  51.06 |  40.43 |   8.51 |
         |   2.23 |   5.65 |   4.30 |
---------+--------+--------+--------+
Total        1078      336       93     1507
            71.53    22.30     6.17   100.00

Frequency Missing = 10
```

Exhibit 10.1

Format of a Basic Crosstab Generated by PROC FREQ

Options for the TABLES Statement The default crosstabulation table output by PROC FREQ contains a great deal of information. When one of the variables is hypothesized to depend on the other, it may contain more information than is needed and may be confusing. When one variable is assumed to be dependent and the other independent, only one set of percentages is generally needed to describe the relationship. If the row variable is the independent variable, usually only the row percentages are needed. Similarly, if the column variable is independent, only the column percentages are needed.

Several options can be used on the TABLES statement to keep unnecessary percentages from appearing in the output. They are NOPERCENT, NOROW, and NOCOL. The NOPERCENT option suppresses the cell percentages of the grand total. The NOROW and NOCOL options suppress the row and column percentages. You may employ the NOPERCENT option and either the NOROW or NOCOL option to leave only the cell frequencies and the relevant row or column percentages.

Two TABLES options specify statistical tests for the null hypothesis that the row and column variables are independent. They provide various measures of the strength of the relationship between the two. The CHISQ option provides Pearson and likelihood ratio chi-square tests for the null hypothesis that the two variables have no relationship, assuming that the data come from a simple random sample. It also provides several measures of the strength of the relationship between the variables that are based on the Pearson chi-square statistic. The MEASURES option provides a number of additional measures of the strength of the relationship between the crosstabulated variables that are not based on chi-square.

Handling Missing Data and Low Cell Counts Many introductory statistics texts warn that chi-square statistics for crosstabulations can be inaccurate when over 20 percent of the expected cell frequencies are less than 5. PROC FREQ automatically checks each table for this condition when it computes chi-square statistics. If it occurs, FREQ prints a warning at the end of the statistics section of the output for the table. If each row and column variable has just two categories and the CHISQ option is used, PROC FREQ automatically calculates the Fisher's Exact chi-square. This gives exact probabilities for significance tests even when expected frequencies are less than 5. If this condition occurs and the number of rows or columns is greater than 2, the EXACT option may be used to request that the Fisher's Exact chi-square be calculated. This calculation is computationally intense. It should be used with great caution, especially for tables where the number of rows or columns exceeds 5 or the number of observations is very large.

Enhancing Table Readability When the variables being crosstabulated are numeric, as in the above example, it is often difficult to interpret the table output by PROC FREQ or to use it directly in a research paper. The crosstabulation output would be clearer if it had mnemonic labels indicating what the values stood for. This can be accomplished by using either of two techniques. You can create character-valued variables from the numeric-valued ones. The other technique is to create custom output formats with PROC FORMAT and add a FORMAT statement after the TABLES statement. Although we'll give you a hint about using PROC FORMAT in a later example, we'll cover the use of PROC FORMAT more thoroughly in Chapter 13. In our next example, we'll create and use some character-valued variables.

Our next example, shown in Exhibit 10.2, creates and uses the character variables `crace` and `ccappun`. This gives the rows and columns of the crosstab readily understandable headings. We also combine the `Black` and `Other` respondents into a single category. In addition, we employ the NOPERCENT, NOCOL,CHISQ,

EXACT, and MEASURES options on the TABLES statement to simplify the crosstabulation output and generate statistics.

Exhibit 10.2

Producing a More Readable Crosstab

Program

```
LIBNAME  data  'location' ;

/* Create character variables in DATA step */
DATA  temp ;
   SET  data.gss1990 ;

   /* Create character-valued var. for punishment   */
   /* Be sure to assign longest value first.         */
   /* Add one blank before Favor and Oppose to make */
   /* values sort in correct order.                  */
   IF        cappun = 8 THEN ccappun = "Don't Know";
   ELSE IF  cappun = 1 THEN ccappun = " Favor";
   ELSE IF  cappun = 2 THEN ccappun = " Oppose";

   /* Create character-valued var. for race    */
   /* Assign longest value first.              */
   /* Add a blank in front of White            */
   IF race = 2  OR race = 3 THEN crace = 'Black & Other';
      ELSE IF race =  1     THEN crace = ' White' ;
RUN;

PROC FREQ  DATA= temp ;

     /* Use options to simplify table and get statistics */
     TABLES  crace*ccappun / NOPERCENT NOCOL CHISQ EXACT MEASURES ;

TITLE1 'Crosstabulation of Race of Respondent and'   ;
TITLE2 'Opinion on Death Penalty Sentence for Murder with' ;
TITLE3 'Tests for and Measures of Association' ;
RUN;
```

Output

```
Crosstabulation of Race of Respondent and
Opinion on Death Penalty Sentence for Murder with
Tests for and Measures of Association

TABLE OF CRACE BY CCAPPUN

CRACE(RACE OF RESPONDENT)
              CCAPPUN(FAVOR OR OPPOSE DEATH PENALTY FOR MURDER)
Frequency    |
Row Pct      |Favor   |Oppose  |Don't   |
             |        |        |Know    |  Total
-------------+--------+--------+--------+
 White       |    947 |    244 |     69 |   1260
             |  75.16 |  19.37 |   5.48 |
-------------+--------+--------+--------+
Black & Other|    131 |     92 |     24 |    247
             |  53.04 |  37.25 |   9.72 |
-------------+--------+--------+--------+
Total            1078      336       93     1507
```

Exhibit 10.2

Producing a More Readable Crosstab (continued)

New Page of Output

```
Crosstabulation of Race of Respondent and
Opinion on Death Penalty Sentence for Murder with
Tests for and Measures of Association

STATISTICS FOR TABLE OF RACE BY CAPPUN

Statistic                        DF       Value         Prob
------------------------------------------------------------
Chi-Square                        2      49.764        0.000
Likelihood Ratio Chi-Square       2      46.213        0.000
Mantel-Haenszel Chi-Square        1      16.570        0.000
Fisher's Exact Test (2-Tail)                        5.80E-11
Phi Coefficient                           0.182
Contingency Coefficient                   0.179
Cramer's V                                0.182


Statistic                                 Value          ASE
------------------------------------------------------------
Gamma                                     0.419        0.053
Kendall's Tau-b                           0.174        0.027
Stuart's Tau-c                            0.120        0.020

Somers' D C|R                             0.220        0.034
Somers' D R|C                             0.138        0.022

Pearson Correlation                       0.105        0.029
Spearman Correlation                      0.178        0.028

Lambda Asymmetric C|R                     0.000        0.000
Lambda Asymmetric R|C                     0.000        0.000
Lambda Symmetric                          0.000        0.000

Uncertainty Coefficient C|R               0.021        0.006
Uncertainty Coefficient R|C               0.034        0.010
Uncertainty Coefficient Symmetric         0.026        0.008

Sample Size = 1507
```

The output in Exhibit 10.2 shows the results of using various TABLES statement options. Note that the rows and columns of the crosstabulation on the first page are clearly labeled and the information in the cells is less cluttered. With the exception of the Fisher's Exact test, the first set of statistics on the second page is provided by the CHISQ option. The degrees of freedom and probabilities are provided for the chi-square statistics. No degrees of freedom or levels of significance are needed for the three measures of association (`Phi Coefficient`, `Contingency Coefficient`, and `Cramer's V`).

The p-value associated with the Fisher's Exact test requested by the EXACT option is so small that it is expressed in scientific notation. The second group of statistics on the second page is the set requested with the MEASURES option. Their values are given under the heading `Value`. Their asymptotic standard errors are provided under the heading `ASE`. A good review of many of these statistics is provided in Gibbons (1993).

Entering Previously Crosstabulated Data

In many research situations, crosstabulations and related statistics are constructed from a dataset that contains individual observations. This is quite common in survey research. However, occasionally you will find that the original data have already been crosstabulated. This is often the case when you wish to analyze a crosstab found in published research or in a government statistical abstract. It is possible to obtain tests and measures of association for existing crosstabulations by reading in the cell frequencies of the table and employing a WEIGHT statement with PROC FREQ.

We illustrate this strategy with the code below. It reads in the cell frequencies of the crosstab presented in Exhibit 10.3. The use of output formats and variable labels is not necessary for producing statistics, but does clearly label the output table. We chose to call the row and column variables `tabrace` and `tabpun`, respectively. The codes for each variable are somewhat arbitrary, but do reflect the ordering of the rows and columns of the crosstabulation in Exhibit 10.3. Thus, the first cell in the first row has the values `tabrace`=1 and `tabpun`=1. The second cell in that row has the values `tabrace`=1 and `tabpun`=2. The first cell in the second row has the values `tabrace`=2 and `tabpun`=1, and so on.

The key to entering data in this form is to have one observation for each cell count of the crosstabulation. This crosstab had six cells, so we entered six cell counts with appropriate values for `tabrace` and `tabpun`.

The SAS dataset will have only six observations. How do six observations become 1507? The answer is that the WEIGHT statement that follows the PROC FREQ statement causes PROC FREQ to include each observation in the crosstabulation `count` number of times. Thus, the first observation is included 947 times, the second, 244 times, and so on. We placed this statement after the PROC FREQ statement, but it could be placed anywhere between the PROC FREQ statement and the RUN statement.

You can compare the cell counts and statistics in Exhibit 10.3 to those in Exhibit 10.2 to assure yourself that this strategy produces bivariate statistics identical to those that were produced when the data for individual respondents were used.

Exhibit 10.3

Creating a Crosstab from Previously Tabulated Data

Program

```
/* Setup formats */
LIBNAME  library  'location' ;
PROC FORMAT LIBRARY=library ;   /* See Chapter 13 for use of FORMAT */
     VALUE  punfmt     1 = 'Favor'
                       2 = 'Oppose'
                       3 = 'Don''t   Know' ;

     VALUE  whtblk     1 = 'White'
                       2 = 'Black & Other'   ;

/* Read in cell counts of the table by row                      */
/* Values for tabrace:  1= White and 2 = Black and Other        */
/* Values for tabpun:   1= Favor  2= Oppose  3= Don't Know      */
DATA  table ;
INPUT    tabrace tabpun count;

      LABEL    tabrace = 'Race of Respondent'
               tabpun  = 'Opinion on Use of Death Penalty' ;

/* Enter counts for six cells by row of table */
DATALINES ;
1 1  947
1 2  244
1 3  69
2 1  131
2 2   92
2 3  24
RUN;

PROC FREQ  DATA=table;
      /* Use WEIGHT statement to get correct cell frequencies */
     WEIGHT count;

     TABLES  tabrace * tabpun / NOPERCENT NOCOL  CHISQ EXACT MEASURES
             ;

     /* Assign formats to variables */
     FORMAT  tabrace whtblk.  tabpun punfmt. ;
RUN;
```

Output

```
TABLE OF TABRACE BY TABPUN

TABRACE(Race of Respondent)
               TABPUN(Opinion on Use of Death Penalty)
Frequency     |
Row Pct       |Favor   |Oppose  |Don't   |
              |        |        |Know    |  Total
--------------+--------+--------+--------+
White         |    947 |    244 |     69 |   1260
              |  75.16 |  19.37 |   5.48 |
--------------+--------+--------+--------+
Black & Other |    131 |     92 |     24 |    247
              |  53.04 |  37.25 |   9.72 |
--------------+--------+--------+--------+
Total             1078      336       93    1507
```

Exhibit 10.3

Creating a Crosstab from Previously Tabulated Data (continued)

New Page of Output

```
STATISTICS FOR TABLE OF TABRACE BY TABPUN

Statistic                            DF      Value        Prob
--------------------------------------------------------------
Chi-Square                            2     49.764       0.000
Likelihood Ratio Chi-Square           2     46.213       0.000
Mantel-Haenszel Chi-Square            1     41.001       0.000
Fisher's Exact Test (2-Tail)                          5.80E-11
Phi Coefficient                              0.182
Contingency Coefficient                      0.179
Cramer's V                                   0.182


Statistic                                    Value         ASE
--------------------------------------------------------------
Gamma                                        0.419       0.053
Kendall's Tau-b                              0.174       0.027
Stuart's Tau-c                               0.120       0.020

Somers' D C|R                                0.220       0.034
Somers' D R|C                                0.138       0.022

Pearson Correlation                          0.165       0.028
Spearman Correlation                         0.178       0.028

Lambda Asymmetric C|R                        0.000       0.000
Lambda Asymmetric R|C                        0.000       0.000
Lambda Symmetric                             0.000       0.000

Uncertainty Coefficient C|R                  0.021       0.006
Uncertainty Coefficient R|C                  0.034       0.010
Uncertainty Coefficient Symmetric            0.026       0.008

Sample Size = 1507
```

Testing for Differences Between Groups on a Rank-Ordered Dependent Variable

Many research situations compare groups of individuals, types of products, or experimental treatments to detect differences among them. Some of these scenarios involve categorical independent variables. When the independent variable is categorical and the dependent variable on which the groups are being compared is rank-ordered, statisticians often recommend the use of nonparametric, or distribution-free, tests. These tests are also often recommended when the normality or equal variance assumptions for a two-group t test or a one-way ANOVA cannot be met. This is because their results do not depend on any parametric assumptions about the underlying distribution of the dependent variable. SAS's PROC NPAR1WAY may be used to perform a wide variety of nonparametric tests.

By default, NPAR1WAY calculates all tests that are appropriate for the number of categories of the independent variable. You can request only the test or tests that you want performed on your data through the use of options on the PROC NPAR1WAY statement. Tests available in PROC NPAR1WAY and the option names for requesting them are shown in Exhibit 10.4.

Exhibit 10.4

PROC NPAR1WAY Statement Options for Nonparametric Tests

Options	Tests
WILCOXON	Wilcoxon or Mann-Whitney U two-group test, Kruskal-Wallis H test for two or more groups
MEDIAN	Two-group median test and K-group Brown-Mood test based on 0,1 median scoring of ranks
SAVAGE	Two-group and K-group tests based on Savage scoring of ranks
VW	Two-group and K-group tests based on Van der Waerden scoring of ranks
EDF	Two-group Kuiper test and two-group and K-group Cramer–von Mises and Kolmogorov-Smirnov tests of differences in distributions

The program in Exhibit 10.5 demonstrates the basic syntax for using PROC NPAR1-WAY for an independent variable with two categories to request the Wilcoxon and Kruskal-Wallis tests. The data are taken from the 1991 General Social Survey. The dependent variable is an occupational prestige score assigned to each respondent's occupation. Note the use of the WILCOXON option. Remember that this option will produce both the Wilcoxon test and the Kruskal-Wallis test for independent variables with two categories, but only the Kruskal-Wallis test for variables with more than two categories.

The CLASS statement is required. Only one variable name may appear on the CLASS statement. As a practical matter, the VAR statement is also required. Omitting it will cause SAS to perform tests for every numeric variable in the input dataset. Placing more than one variable name on the VAR statement will produce separate tests for each dependent variable.

Exhibit 10.5

PROC NPAR1WAY Used with WILCOXON Option

Program

```
LIBNAME  data     'location' ;

/* Use WILCOXON option to get only Wilcoxon */
/* and Kruskal-Wallis tests.                */
PROC NPAR1WAY  DATA=data.tgss1991  WILCOXON ;

     /* Name of only one independent variable    */
     CLASS  sex ;

     /* Name of one or more dependent variables */
     VAR    prestg80 ;

     /* Optional variable label for dependent var.    */
     LABEL   prestg80 = 'Occupational Prestige' ;


TITLE1  'Nonparametric Tests for Difference in' ;
TITLE2  'Occupational Prestige Ranking for'     ;
TITLE3  'Male and Female Respondents'           ;
RUN;
```

Output

```
Nonparametric Tests for Difference in
Occupational Prestige Ranking for
Male and Female Respondents

 N P A R 1 W A Y  P R O C E D U R E

Wilcoxon Scores (Rank Sums) for Variable PRESTG80
Classified by Variable SEX


                          Sum of     Expected      Std Dev          Mean
   SEX          N         Scores     Under H0      Under H0         Score

    1         797     546518.500   565471.500   7645.16265    685.719573
    2         621     459552.500   440599.500   7645.16265    740.020129

Average Scores were used for Ties
Wilcoxon 2-Sample Test (Normal Approximation)
(with Continuity Correction of .5)

S=   459553         Z=  2.47902          Prob > |Z| =    0.0132

T-Test approx. Significance =     0.0133

Kruskal-Wallis Test (Chi-Square Approximation)
CHISQ=  6.1459        DF=  1        Prob > CHISQ=      0.0132
```

As you can see in Exhibit 10.5, PROC NPAR1WAY produces a table and associated sets of significance tests. Reading the columns of the output table from left to right, the table lists the values of the categorical independent variable under its variable name (`sex`), the number of observations without missing values in each group under the heading `N`, the actual sum of the rank values for each group, the sum of the ranks that would be expected if there were no difference between the groups (under the heading `Expected Under H0`), the standard deviation of the expected sums, and the mean value of the ranks assigned to the prestige scores.

The significance tests assess whether the difference in the rank sums could be due to chance. When the number of groups is two, NPAR1WAY first gives the normal approximation test for the Wilcoxon statistic, based on a Z statistic. In this example, the value of Z is 2.47902 and the associated probability is .0132. It then displays the significance level of the two-group T statistic proposed by Mann and Whitney, which in this case is .0133. NPAR1WAY always outputs the chi-square approximation for the significance of the Kruskal-Wallis test when the WILCOXON option is used. Here, the chi-square value with 1 degree of freedom is 6.1459, with an associated probability of .0132.

If you wish to improve your undertanding of these and of other nonparametric tests produced by PROC NPAR1WAY in general, see Daniel (1990) or Mosteller and Rourke (1973). For a discussion of nonparametric tests as robust alternatives to conventional ANOVA, see Maxwell and Delaney (1990), Chapter 15.

Logistic Regression: Estimating a Prediction Equation for a Dichotomous Dependent Variable

There are many situations in which a researcher wishes to develop a simple or multiple regression model for predicting a dependent variable that has two categories. Examples of such variables are a respondent's voting Republican versus Democratic and contracting a disease versus staying well. One commonly used statistical technique for estimating such models is logistic regression. Logistic regression models can be extended to dependent variables with more than two categories. However, this is beyond.the scope of this text.

SAS provides a number of PROCs that may be employed for performing logistic regression analysis. We will concentrate on two that we feel meet most basic needs in a relatively straightforward manner — PROC LOGISTIC and PROC CATMOD. The choice between them is governed mainly by whether the independent variables are primarily categorical or continuous in nature. PROC LOGISTIC, like PROC REG, is most useful when the independent variables are all or mostly continuous. PROC CATMOD, like PROC GLM, is most useful when all of the independent variables are categorical.

PROC LOGISTIC: Estimating Logistic Regression Models with Continuous Independent Variables

The code in Exhibit 10.6 illustrates the basic use of PROC LOGISTIC with no options to estimate a simple logistic regression model with two continuous independent variables. The data, taken from the 1991 General Social Survey, are for respondents who reported voting for either Dukakis or Bush in the 1988 presidential election. The variable `pres88` is coded 1 if the respondent reported voting for Dukakis and 2 for voting for Bush. An estimate of the respondent's family income and the number of years the father went to school are used as predictors.

Exhibit 10.6

Using PROC LOGISTIC with No Options

Program

```
LIBNAME   data   'location' ;

/* Presidential Vote: 1 = Dukakis  2 = Bush */
PROC LOGISTIC  DATA=data.presvote              ;
     MODEL  pres88 =  income91   paeduc ;

TITLE1 'Predicting Presidential Vote for Dukakis in 1988 from'   ;
TITLE2 'Family Income and Number of Years of Father''s Education' ;
RUN;
```

Output

```
Predicting Presidential Vote for Dukakis in 1988 from
Family Income and Number of Years of Father's Education

The LOGISTIC Procedure

Data Set: DATA.PRESVOTE
Response Variable: PRES88     VOTE FOR DUKAKIS OR BUSH
Response Levels: 2
Number of Observations: 752
Link Function: Logit


     Response Profile

Ordered
  Value  PRES88      Count

      1       1        296
      2       2        456

WARNING: 718 observation(s) were deleted due to missing values for the
         response or explanatory variables.
```

Exhibit 10.6

Using PROC LOGISTIC with No Options (continued)

```
                  Criteria for Assessing Model Fit

                              Intercept
                Intercept        and
Criterion         Only       Covariates   Chi-Square for Covariates

AIC             109.189      1001.575          .
SC              1014.812     1015.443          .
-2 LOG L        1008.189     995.575       12.614 with 2 DF (p=0.0018)
Score               .            .          12.660 with 2 DF (p=0.0018)
 Analysis of Maximum Likelihood Estimates

               Parameter  Standard      Wald        Pr >      Standardized
Variable  DF    Estimate    Error    Chi-Square  Chi-Square     Estimate

INTERCPT  1     -0.3887    0.1392       7.7984      0.0052            .
INCOME    1    -0.00505   0.00249       4.1150      0.0425     -0.087651
PAEDUC    1     0.00652   0.00242       7.2674      0.0070      0.109808
```

New Page of Output

```
Predicting Presidential Vote for Dukakis in 1988 from
Family Income and Number of Years of Father's Education

The LOGISTIC Procedure

         Analysis of Maximum Likelihood Estimates

             Odds     Variable
Variable     Ratio      Label

INTERCPT     0.678    Intercept
INCOME       0.995    Total Family Income
PAEDUC       1.007    HIGHEST YEAR SCHOOL COMPLETED, FATHER


Association of Predicted Probabilities and Observed Responses

 Concordant = 55.9%           Somers' D = 0.137
 Discordant = 42.2%           Gamma     = 0.140
 Tied       =  2.0%           Tau-a     = 0.065
 (134976 pairs)               c         = 0.568
```

LOGISTIC Syntax The MODEL statement, placed after the PROC LOGISTIC statement, lists the dependent and independent variables for the equation. A single dependent variable name is placed to the left of an equal sign. The dependent variable may be either character or numeric. The SAS variable names of the independent variables are listed to the right of the equal sign. All independent variables must be numeric.

PROC LOGISTIC has several options that are useful for basic logistic regression analysis. Perhaps the most useful is the DESCENDING option, which can be added to the PROC LOGISTIC statement. This option instructs PROC LOGISTIC to form predicted logits by comparing the higher value of the dependent variable to the lower. It does this by making the high data value the first ordered value. Using this option can eliminate the confusion we talked about earlier, because the signs of the coefficients of the resulting logit prediction equation make more intuitive sense.

Two other options can be employed to obtain a useful fit statistic and regression diagnostics. Both of these options may appear on the MODEL statement. They are placed after the name of the last independent variable and separated from it by a slash. The Hosmer-Lemeshow goodness-of-fit test may be obtained by using the LACKFIT option (Hosmer and Lemeshow, 1989). Diagnostic statistics for identifying poorly predicted or very influential observations, proposed by Pregibon (1981), may be obtained by employing the INFLUENCE option. You should be aware that using this option can create a great number of pages of output for datasets with large numbers of observations.

Interpreting LOGISTIC Output Review the output in Exhibit 10.6. The first block of information in the upper left corner of the first page lists the name of the SAS input dataset (`DATA.PRESVOTE`), the response or dependent variable name (`PRES88`), the number of categories of the dependent variable (`Response Levels: 2`), and the number of observations from the dataset used in the analysis. The specification of `Logit` as the `Link Function` indicates that a logistic regression was performed. PROC LOGISTIC is capable of estimating models with other link functions. It performs logistic regression by default.

The information listed under the heading `Response Profile` gives a frequency count for each value of the dependent variable and issues a warning about the number of observations in the input dataset excluded from the analysis. PROC LOGISTIC uses a list-wise exclusion of observations. This means that if any independent or dependent variable in an observation has a missing value, that observation is not used for estimating the regression model.

The really important thing to note in the Response Profile section is the `Ordered Value` assigned to the actual values of the dependent variable. The *ordered* values, not the *actual* variable values, determine the form of the logit predicted by the logistic regression equation. PROC LOGISTIC always forms the predicted logit for a dichotomous dependent variable as

$$\text{Log}_e[(\text{Probability of Ordered Value} = 1) \div (\text{Probability of Ordered Value} = 2)]$$

which, for our example data, compares the probability of voting for Dukakis to that of voting for Bush.

The fact that PROC LOGISTIC generally forms the logit for predicting the lower-ordered value of a dependent variable rather than that for the higher-ordered value creates great confusion for many beginning users of this PROC. When they examine the signs of the coefficients of the prediction equation, the signs appear to be the opposite of what they expect. This is because PROC LOGISTIC is comparing the probability of the lower value to that of the higher, rather than that of the higher to that of the lower (as is most often done in statistical texts). In our next example, we will show you how to use the DESCENDING option to tell PROC LOGISTIC to form the logit in the more conventional manner.

The next section of the output on the first page gives statistics for assessing the fit of the model to the data. Values are given for four of these statistics for two logistic models — one that includes only an intercept term and one that includes an intercept and the predictor variables, or *covariates*. The values of the statistics for these two models are listed under the column headings `Intercept Only` and `Intercept and Covariates`. The fit statistics include the Akaike Information Criterion, labeled `AIC`, and the Schwartz Criterion, labelled `SC`. These two statistics are used for comparing models; thus, they are not useful for measuring the contribution of the independent variables to predicting the dependent variable. No value is ever given for them in the `Chi-square for Covariates` column. The difference in the `-2` Log Likelihood statistics for the two models yields a chi-square statistic that can be used to test for the overall significance of the predictors. The `Score` statistic has a chi-square distribution, but is not based on a comparison between models. Thus, the chi-square value and probability are shown, but no values are placed in the two model comparison columns.

The next section of the output, `Analysis of Maximum Likelihood Estimates`, presents the maximum likelihood estimates of the coefficients of the logistic prediction equation and their standard errors under the heading `Parameter Estimate`. A Wald chi-square and associated probability for testing the difference of each coefficient from zero is also displayed. The standardized coefficients and odds ratios are also given. The odds ratios are computed for a one-unit increase in the independent variable by raising the base of the natural logarithm to the power of the coefficient.

One way to assess the predictive efficacy of a logistic model is to examine the degree of association between the predicted probabilities for the lower-ordered value based on the data values for the predictor variables and the estimated coefficients. This can be accomplished by looking at the section of the output on the second page labeled `Association of Predicted Probabilities and Observed Responses`. Four rank correlation statistics, calculated from the numbers of concordant and discordant pairs of observations, are provided. The better-known statistics are Somers' D, Goodman and Kruskal's Gamma, and Kendall's tau-a. The formula for the less well-known c is

```
c = (n of concordant pairs + .5(n of tied pairs))
      ÷ total pairs
```

For assistance in interpreting these statistics as well as other measures of predictive efficacy, see Aldrich and Nelson (1984), Hosmer and Lemeshow (1989), or Demaris (1992).

Because the output for the INFLUENCE option for our presidential vote data would be voluminous, we only demonstrate the use of the DESCENDING and LACKFIT options in the program in Exhibit 10.7. Note that because we used the DESCENDING option, the TITLE1 statement has been altered to reflect the fact that the predicted logit is voting for Bush.

Exhibit 10.7

PROC LOGISTIC Using the DESCENDING and LACKFIT Options

Program

```
LIBNAME  data    'location' ;
LIBNAME library 'c:\sasbook\datasets' ;

/* Presidential Vote: 1 = Dukakis  2 = Bush */
/* Note use of the DESCENDING option        */
PROC LOGISTIC  DATA=data.presvote  DESCENDING ;

     /*  Request Hosmer-Lemeshow Test  */
     MODEL  pres88 =  income91   paeduc  /  LACKFIT ;

TITLE1 'Predicting Presidential Vote for Bush in 1988 from';
TITLE2 'Family Income and Number of Years of Father''s Education';
RUN;
```

Output

```
Predicting Presidential Vote for Bush in 1988 from
Family Income and Number of Years of Father's Education

The LOGISTIC Procedure

Data Set: DATA.PRESVOTE
Response Variable: PRES88    VOTE FOR DUKAKIS OR BUSH
Response Levels: 2
Number of Observations: 752
Link Function: Logit


     Response Profile

Ordered
  Value  PRES88     Count

      1       2       456
      2       1       296

WARNING: 718 observation(s) were deleted due to missing values for the
         response or explanatory variables.
```

Exhibit 10.7

PROC LOGISTIC Using the DESCENDING and LACKFIT Options (continued)

```
                    Criteria for Assessing Model Fit

                                 Intercept
                    Intercept       and
Criterion             Only       Covariates    Chi-Square for Covariates

AIC                 1010.189      1001.575          .
SC                  1014.812      1015.443          .
-2 LOG L            1008.189       995.575       12.614 with 2 DF (p=0.0018)
Score                   .              .          12.660 with 2 DF (p=0.0018)
Analysis of Maximum Likelihood Estimates

                  Parameter  Standard      Wald          Pr >      Standardized
Variable   DF     Estimate     Error    Chi-Square   Chi-Square      Estimate

INTERCPT    1       0.3887     0.1392       7.7984       0.0052            .
INCOME      1      0.00505    0.00249       4.1150       0.0425      0.087651
PAEDUC      1     -0.00652    0.00242       7.2674       0.0070     -0.109808
```

New Page of Output

```
Predicting Presidential Vote for Bush in 1988 from
Family Income and Number of Years of Father's Education

The LOGISTIC Procedure

          Analysis of Maximum Likelihood Estimates

              Odds      Variable
Variable      Ratio       Label

INTERCPT      1.475     Intercept
INCOME        1.005     Total Family Income
PAEDUC        0.993     HIGHEST YEAR SCHOOL COMPLETED, FATHER

Association of Predicted Probabilities and Observed Responses

 Concordant = 55.9%           Somers' D = 0.137
 Discordant = 42.2%           Gamma     = 0.140
 Tied       =  2.0%           Tau-a     = 0.065
 (134976 pairs)               c         = 0.568

            Hosmer and Lemeshow Goodness-of-Fit Test

                               PRES88 = 2              PRES88 = 1
                          --------------------    --------------------
 Group        Total       Observed    Expected    Observed    Expected

    1           75              31       34.61          44       40.39
    2           75              44       41.40          31       33.60
    3           75              43       44.52          32       30.48
    4           75              48       45.13          27       29.87
    5           75              43       45.76          32       29.24
    6           75              50       46.39          25       28.61
    7           75              50       47.08          25       27.92
    8           75              50       47.89          25       27.11
    9           75              46       48.98          29       26.02
   10           77              51       54.25          26       22.75

Goodness-of-fit Statistic = 4.7337 with 8 DF (p=0.7856)
```

Compare the output in Exhibit 10.7 with that in Exhibit 10.6. Look at the ordered values for `pres88` = 2 in the `Response Profile` section of the output on the first page. Use of the DESCENDING option in this example made voting for Bush the first ordered value rather than the second. Notice that all the fit statistics in the `Criteria for Assessing Model Fit` section are identical. Notice also that the parameter estimates in Exhibit 10.7 have the same absolute values as those in Exhibit 10.6, but that they have opposite signs. The standard errors, Wald chi-square values and associated probabilities, and rank correlations between the predicted probabilities and the observed responses are also identical.

The Hosmer-Lemeshow test for the goodness of fit of the model to the data is shown at the bottom of the second page. The table gives the observed and expected frequencies of voting for Dukakis and Bush across the range of predicted probability of voting for Bush. The predicted probabilities are grouped into the 10 categories listed under the heading `Group`. The `Goodness-of-fit Statistic` at the bottom of the table is a Pearson chi-square calculated from the observed and expected frequencies given in the table. (See Hosmer and Lemeshow (1989) for details about the calculation and interpretation of this test.)

PROC CATMOD: Estimating Logistic Regression Models with Categorical Independent Variables

Either PROC LOGISTIC or PROC CATMOD may be employed to estimate logistic regression models with categorical independent variables. PROC CATMOD is more efficient from the user's perspective, because it automatically creates the necessary dummy variables for the independent variables. It also has a convenient syntax (similar to that of PROC GLM) for specifying main effect and interaction terms. In addition, analyses may easily be performed for datasets containing individual observations or for previously crosstabulated data.

Only one MODEL statement may be used with a PROC CATMOD statement.The syntax for the specification of the effects in your prediction equation is identical to that of the preceding chapter's MODEL statement in PROC GLM. Individual variable names are treated as main effects; variable names linked with asterisks, as interactions. The vertical bar between variable names may also be used to include main and interaction effects.

The code in Exhibit 10.8 uses data from Higgins and Koch (1977). It reads in previously crosstabulated data and estimates a logistic regression equation for predicting the presence of brown lung disease (byssinosis) in a sample of textile mill workers. The independent variables are presence or absence of dusty working conditions, number of years employed, and whether the individual was a smoker or a nonsmoker. The count for each cell of the $2 \times 3 \times 2 \times 2$ crosstabulation of the original data on workers is entered as a separate observation in the data following the DATALINES statement.

The WEIGHT statement following the PROC CATMOD statement is used to create the proper number of observations for each cell of the crosstabulation analyzed by PROC CATMOD. If data for individual workers, rather than the crosstabulated data, were read in, the WEIGHT statement would not be necessary. See "Entering Previously Crosstabulated Data," earlier in this chapter, for a discussion of the WEIGHT statement.

Exhibit 10.8

Using PROC CATMOD for Previously Tablulated Data

Program

```
/* Read in 2 x 3 x 2 x 2 crosstabulation  */
DATA  lung ;
      /* Declare lengths of character vars (see Chapter 14) */
      LENGTH dust $ 9   yrsemp $ 5   smoker $ 3  byssin $ 3 ;
      INPUT  dust $ 1-9  @14 yrsemp $CHAR5.   @21smoker $CHAR3.
             @26 byssin $CHAR3.   count 30-33 ;
      LABEL  dust   = 'Workplace Dust Conditions'
             yrsemp = 'Years Employed in Mill'
             smoker = 'Smoker'
             byssin = 'Byssinotic Lungs'  ;
DATALINES;
Dusty         <10   No   Yes    7
Dusty         <10   No   No   119
Dusty         <10   Yes  Yes   30
Dusty         <10   Yes  No   203
Dusty        10-20  No   Yes    3
Dusty        10-20  No   No    17
Dusty        10-20  Yes  Yes   16
Dusty        10-20  Yes  No    51
Dusty        >20    No   Yes    8
Dusty        >20    No   No    64
Dusty        >20    Yes  Yes   41
Dusty        >20    Yes  No   110
Not Dusty     <10   No   Yes   12
Not Dusty     <10   No   No  1004
Not Dusty     <10   Yes  Yes   14
Not Dusty     <10   Yes  No  1340
Not Dusty    10-20  No   Yes    2
Not Dusty    10-20  No   No   209
Not Dusty    10-20  Yes  Yes    5
Not Dusty    10-20  Yes  No   409
Not Dusty    >20    No   Yes    8
Not Dusty    >20    No   No   777
Not Dusty    >20    Yes  Yes   19
Not Dusty    >20    Yes  No   951

PROC CATMOD ;
   /* Use WEIGHT statement to get correct cell counts. This
      statement isn't necessary if reading individual obs */
   WEIGHT count;
   /* Model contains 3 IV's and interactions of dust with other two */
   MODEL  byssin =  dust yrsemp smoker  dust*yrsemp dust*smoker ;
TITLE1 'Predicting NO Byssinosis from Dust Conditions,';
TITLE2 'Years of Employment and Smoking Status'        ;
RUN;
```

Exhibit 10.8

Using PROC CATMOD for Previously Tablulated Data (continued)

Output

```
Predicting NO Byssinosis from Dust Conditions,
Years of Employment and Smoking Status

CATMOD PROCEDURE

Response: BYSSIN                    Response Levels (R)=     2
Weight Variable: COUNT              Populations     (S)=    12
Data Set: LUNG                      Total Frequency (N)=  5419
Frequency Missing: 0                Observations  (Obs)=    24


            POPULATION PROFILES
                                          Sample
Sample     DUST       YRSEMP  SMOKER       Size
-------------------------------------------------
    1   Dusty          <10     No           126
    2   Dusty          <10     Yes          233
    3   Dusty         10-20    No            20
    4   Dusty         10-20    Yes           67
    5   Dusty         >20      No            72
    6   Dusty         >20      Yes          151
    7   Not Dusty      <10     No          1016
    8   Not Dusty      <10     Yes         1354
    9   Not Dusty     10-20    No           211
   10   Not Dusty     10-20    Yes          414
   11   Not Dusty     >20      No           785
   12   Not Dusty     >20      Yes          970


RESPONSE PROFILES

Response  BYSSIN
----------------
    1       No
    2       Yes
```

New Page of Output

```
Predicting NO Byssinosis from Dust Conditions,
Years of Employment and Smoking Status

                  MAXIMUM-LIKELIHOOD ANALYSIS

                                                         Parameter
                   Sub         -2 Log     Convergence    Estimates
Iteration      Iteration     Likelihood    Criterion         1
-------------------------------------------------------------------
     0             0         7512.3291       1.0000          0
     1             0         2067.7441       0.7248       1.6561
     2             0         1380.7121       0.3323       2.3472
     3             0          1219.878       0.1165       2.7873
     4             0          1193.676       0.0215       3.0294
     5             0          1192.101     0.001319       3.1065
     6             0         1192.0895    9.6765E-6       3.1136
     7             0         1192.0895    8.329E-10       3.1137
```

Exhibit 10.8

Using PROC CATMOD for Previously Tablulated Data (continued)

```
Parameter Estimates
Iteration          2            3            4            5
-------------------------------------------------------------
    0              0            0            0            0
    1        -0.2950       0.1473      -0.0581       0.1071
    2        -0.6341       0.2577      -0.0962       0.2046
    3        -0.9801       0.3029      -0.0964       0.2570
    4        -1.2142       0.3247      -0.0770       0.2824
    5        -1.2912       0.3382      -0.0636       0.2980
    6        -1.2983       0.3403      -0.0617       0.3006
    7        -1.2984       0.3403      -0.0616       0.3007


                     Parameter Estimates
Iteration          6            7            8
-----------------------------------------------
    0              0            0            0
    1         0.1414      -0.0642       0.1009
    2         0.2397      -0.1149       0.1856
    3         0.2586      -0.1420       0.2098
    4         0.2407      -0.1631       0.1916
    5         0.2272      -0.1765       0.1760
    6         0.2251      -0.1784       0.1734
    7         0.2251      -0.1784       0.1734
```

New Page of Output

```
Predicting NO Byssinosis from Dust Conditions,
Years of Employment and Smoking Status

MAXIMUM-LIKELIHOOD ANALYSIS-OF-VARIANCE TABLE

Source                  DF    Chi-Square       Prob
----------------------------------------------------
INTERCEPT                1        807.88     0.0000
DUST                     1        140.48     0.0000
YRSEMP                   2         11.55     0.0031
SMOKER                   1          9.68     0.0019
DUST*YRSEMP              2          3.32     0.1900
DUST*SMOKER              1          3.22     0.0727

LIKELIHOOD RATIO         4          2.33     0.6757


ANALYSIS OF MAXIMUM-LIKELIHOOD ESTIMATES

                                          Standard      Chi-
Effect             Parameter   Estimate      Error    Square     Prob
-----------------------------------------------------------------------
INTERCEPT                  1     3.1137     0.1095    807.88   0.0000
DUST                       2    -1.2984     0.1095    140.48   0.0000
YRSEMP                     3     0.3403     0.1243      7.50   0.0062
                           4    -0.0616     0.1659      0.14   0.7102
SMOKER                     5     0.3007     0.0966      9.68   0.0019
DUST*YRSEMP                6     0.2251     0.1243      3.28   0.0701
                           7    -0.1784     0.1659      1.16   0.2821
DUST*SMOKER                8     0.1734     0.0966      3.22   0.0727
```

Interpreting CATMOD Output Look at the output in Exhibit 10.8. The names of the dependent, or `Response`, variable, the weighting variable, and the SAS dataset are given in the upper left corner of the first page, along with the number of observations with missing values. In our example, there are no observations with missing values. In the top right corner PROC CATMOD displays the number of `Response Levels` of the dependent variable. The number of `Populations` is the number of cells in the cross-classification of the independent variables, which in this instance is 12. `Total Frequency` is the total of the weighted number of observations in the crosstabulation. The number of `Observations`, on the other hand, is the number of observations in the SAS dataset, which for this table is 24.

The section of the output labeled `POPULATION PROFILES` gives the cross-classification of the independent variables and the total number of observations for each combination of values, listed under the heading `Sample Size`.

The information under the heading `RESPONSE PROFILES` is quite important. It tells you how the logit is formed for the prediction equation. Like PROC LOGISTIC, CATMOD compares the dependent variable's low value to the high value. However, instead of labeling the values as ordered values, it simply calls them `Response 1` and `2`. PROC CATMOD, therefore, forms the predicted logit as

```
Loge[(Probability of Ordered Value = 1)
   ÷ (Probability of Ordered Value = 2)]
```

The `RESPONSE PROFILES` section tells you how the response levels match up with the codes for the dependent variable in the model. Notice that for our example, response level 1 corresponds to *no* brown lung disease.

The `MAXIMUM-LIKELIHOOD ANALYSIS` section at the top of the second page gives technical details about the iterative algorithm used to calculate the maximum-likelihood estimates of the model parameters. It is usually not necessary to examine this output in great detail. The value of -2 log likelihood for the model can be found in the last row of the column, labeled `-2 Log Likelihood`. For this example, the value is 1192.0895.

The section labeled `MAXIMUM-LIKELIHOOD ANALYSIS-OF-VARIANCE TABLE`, on the third page, does contain important information about tests for the overall effect of the terms in the the model. These tests are analogous to those found in a conventional ANOVA table for a balanced ANOVA design since they test for the overall significance of each effect listed on the MODEL statement, controlled for all other effects. They are based on Wald chi-squares.

The likelihood ratio test is printed at the bottom of the ANOVA table. It compares the specified model with the best possible, or "saturated," model for the data. It is not the same test as the chi-square for covariates test given in PROC LOGISTIC. You can construct the latter test by running an additional PROC CATMOD with a MODEL statement of the form

```
MODEL depvar = ;
```

where *depvar* is the name of your dependent variable. The MAXIMUM-LIKELIHOOD ANALYSIS output section for this model gives the value of −2 log likelihood for the intercept-only model. To obtain a chi-square value for a test of the covariates in the model, subtract the −2 log likelihood value for your hypothesized model from that for the intercept-only model. The number of degrees of freedom will be equal to the number of independent variables.

The final section of the output on the third page, entitled ANALYSIS OF MAXIMUM-LIKELIHOOD ESTIMATES, gives the estimates of the coefficients of the prediction equation based on the dummy variables created for each categorical variable and the appropriate cross-product dummy variables for the interactions. PROC CATMOD uses a 1,0,−1 coding for its dummy variables rather than a 0,1 coding. This means that odds ratios cannot be obtained merely by raising the base of the natural log to the power of the value of the estimate (i.e., Odds Ratio = e^b). Rather, the correct procedure is to raise e to the power of twice the value of the estimate provided by PROC CATMOD (i.e., Odds Ratio = e^{2b}).

If you'd like to know more about how to implement advanced logistic regression techniques with SAS, we suggest you read Hosmer and Lemeshow's *Applied Logistic Regression Analysis* and SAS Institute's *Logistic Regression Examples Using the SAS® System, Version 6, First Edition*. The complete references to these works may be found in the "References" section at the end of this chapter.

Recap

Syntax Summary

Here's a summary of the syntax of the statistical procedures needed to implement the statistical techniques you've learned about in this chapter. (See the syntax description in Chapter 1 for an explanation of the notation.)

PROC FREQ for producing crosstabulations and tests and measures of association:

```
PROC FREQ  DATA=SAS-data-set ;
<WEIGHT  weightvar ; >*
TABLES    rowvar * colvar |
   rowvar * (colvar1 <colvar2 ...>) |
   <rowvar1 rowvar2 ...> * colvar
   </ <NOCOL>  <NOROW> <NOPERCENT>
      <CHISQ>   <EXACT>   <MEASURES> >;
```

*Note: For use with tabular data.

PROC NPAR1WAY for testing differences between groups:

```
PROC NPAR1WAY  DATA=SAS-data-set  <EDF> <MEDIAN>
<SAVAGE>  <WILCOXON>  <VW> ;
CLASS groupvar  ;
VAR varname1 <varname2 ...> ;
```

PROC LOGISTIC for predicting a dichotomous dependent variable:

```
PROC LOGISTIC DATA=SAS-data-set <DESCENDING>;
MODEL depvar = xvar1 <xvar2 ...>
      </  <LACKFIT> > ;
```

PROC CATMOD for predicting a dichotomous dependent variable. See the discussion of PROC GLM in Chapter 9 for the syntax of main, interaction, and nested effects.

```
PROC CATMOD  DATA=SAS-data-set ;
<WEIGHT  weightvar ;>*
MODEL   depvar =  xvar1 <xvar2 ...> ;
```

*Note: For use with tabular data.

Content Review

In this chapter, we have discussed the use of PROC FREQ, PROC NPAR1WAY, PROC LOGISTIC, and PROC CATMOD to investigate relationships between categorical or rank-ordered dependent variables and independent variables that may be categorical or continuous. We have not been able to cover every possibility, but have provided you with a foundation for using SAS to obtain most commonly used bivariate tests and measures of association and have introduced you to techniques for estimating and assessing the fit of logistic regression models.

References

Aldrich, John H. and Forrest D. Nelson, *Linear Probability, Logit and Probit Models* (Sage University Paper series on Quantitive Applications in the Social Sciences 007-045). Newbury Park, CA: Sage Publications, 1984.

Daniel, Wayne W., *Applied Nonparametric Statistics, Second Edition*. Boston: PWS-Kent, 1990.

Demaris, Alfred, *Logit Modeling: Practical Applications* (Sage University Paper series on Quantitative Applications in the Social Sciences 007-86). Newbury Park, CA: Sage Publications, 1992.

Gibbons, Jean D., *Nonparametric Measures of Association*. Newbury Park, CA: Sage Publications, 1993.

Higgins, James and Gary G. Koch, "Variable Selection and Generalized Chi-Square Analysis of Categorical Data Applied to a Large Cross-Sectional Occupational Health Survey," *International Statistical Review* 45 (1977), 51–62.

Hosmer, D. W. and S. Lemeshow, *Applied Logistic Regression Analysis.* New York: John Wiley and Sons, 1989.

Maxwell, Scott E. and Harold D. Delaney, *Designing Experiments and Analyzing Data: A Model Comparison Perspective*. Belmont, CA: Wadsworth Publishing Co., 1990.

Mosteller, Frederick and Robert E. K. Rourke, *Sturdy Statistics*. Reading, MA: Addison-Wesley Publishing Co., 1973.

Pregibon, D., "Logistic Regression Diagnostics," *Annals of Statistics* 9, 705–724.

SAS Institute Inc., *Logistic Regression Examples Using the SAS® System, Version 6, First Edition*. Cary, NC: SAS Institute Inc., 1995.

More About SAS Programming

The preceding chapters have probably given you more than a hint that SAS is not only useful for computing statistics but is also a powerful programming language. There is *much* more to SAS than what has been described in this book. This chapter will discuss some other features of the SAS System that are of particular interest to beginners. While reading it, be aware that if you don't see something you'd like to do with the SAS language, it probably *can* be done, but we didn't have the room to put it in the current volume. (See Appendix B for a description of more advanced SAS programming texts.)

A requirement of SAS users at all levels of expertise is to occasionally exert control over various aesthetic and computational features. The first section of this chapter discusses the use of system and dataset options that give you the ability to control features such as the number of observations used by a procedure, the display of the date in output files, and the centering of output.

The next section touches on a few new features of the DATA step. A means to group logically related statements is presented, followed by *functions*. These are computational shortcuts you can use in calculations.

Finally, the last section discusses some of the features that are at your disposal when you are handling date-oriented data. It presents some input and display formats and shows how dates can be used easily in calculations.

Options

The DATA steps and PROCs that you run do not exist in a vacuum. The SAS System makes many assumptions about the environment in which your program runs. These assumptions are commonly referred to as default settings, system settings, or simply defaults. They are usually acceptable, but must sometimes be changed to suit your needs. This section explains how to view and change system options.

Identifying System Settings

Changing system settings is a very straightforward process. First, though, you must know the name, meaning, and default value of an option. This is easily done using either of two techniques. First, for interactive users of SAS, is the Options window in the Display Manager (the Display Manager is discussed in detail in Appendix A). Enter `options` in the command line and press Enter or click on `Global` | `Options` | `Global Options`, and you will see a screen resembling that shown in Exhibit 11.1.

Exhibit 11.1

OPTIONS Window (SAS System for Windows, Version 6.10)

```
SAS - [OPTIONS]
File  Edit  View  Globals  Options  Window  Help

Select option to toggle option value:

[X] BYERR          [X] BYLINE       [ ] CAPS        [ ] CARDIMAGE
[X] CENTER         [ ] CHARCODE     [X] CLEANUP     [ ] CMDMAC
[X] DATE           [ ] DBCS         [ ] DETAILS     [X] DSNFERR
[ ] ERRORABEND     [X] FMTERR       [X] GWINDOW     [ ] IMPLMAC

Enter new option value to update:

BUFNO          1
BUFSIZE        0
CBUFNO         0
COMPRESS       NO
DEVICE
DFLANG         ENGLISH
DKRICOND       ERROR
DKROCOND       WARN
ERRORCHECK     NORMAL
ERRORS         20
FIRSTOBS       1
FMTSEARCH
FORMCHAR       |----|+|---+=|-/\<>*
FORMDLIM
FORMS          DEFAULT
INVALIDDATA    .
LINESIZE       80
MAPS           MAPS
MISSING        .
MSGLEVEL       N

NOTE: At left side.                         C:\SAS610
```

The format will vary depending on your version of SAS and the hardware you're running it on. The effect, however, is the same in all cases. An option is displayed along with its current value. Some options are "yes/no" or "on/off" in nature (the CENTER option is checked in the exhibit, indicating that it is in effect). You can change the value by clicking on the check box or by entering a blank or nonblank character.

Other options require a value. The LINESIZE option in the exhibit has a default value of 80. Its value may be changed simply by entering a new value.

The other way to examine system settings is the OPTIONS procedure. Its simplest syntax is

```
PROC OPTIONS;
```

The procedure will display the name of the option, its current value, and a brief explanation of what the option controls. The explanatory portion of the output makes it ideal for new and occasional users of SAS, since it is a terse but exhaustive list of every setting under your control.

If you can do without the explanation, you have two ways to limit the output produced by the procedure. The form shown below omits the explanation and presents the options and settings in a compact layout.

```
PROC OPTIONS SHORT;
```

If you already know the name of an option and just want to see its current setting, use the following syntax:

```
PROC OPTIONS OPTION=option_name;
```

In the above, *option_name* is the name of a SAS System option. The procedure will display only the current setting of this option. Only one *option_name* may be specified.

Frequently Used Options

As noted above, you may change the value of one or more options by using the Options window in the Display Manager. However, the OPTIONS statement, described in this section, is often a better way to change values. Because it is actually part of your program, you can be sure the value will be set correctly. The syntax of the OPTIONS statement is demonstrated in these examples:

```
OPTIONS ls=78;
OPTIONS NOCENTER NODATE PS=60 LS=78;
```

Keep in mind a few things about the statement and the way you specify values. First, once set, an option remains in effect until the end of the program or until it is reset with another OPTIONS statement. Second, any number of options may be specified. Third, options of the "on/off" variety are prefixed by NO if they are turned off. Otherwise, the option name itself indicates that the option is turned on.

Finally, remember that any number of OPTIONS statements may appear anywhere in your SAS program. One statement may turn a feature on and a later one may turn the feature off. Some frequently used options and their possible settings are presented in alphabetical order in Exhibit 11.2.

Exhibit 11.2

SAS System Options

Option	Comments
CENTER	Centers output horizontally on the page. Turn off by entering NOCENTER.
DATE	Prints the current date at the top of each page of procedure output. Turn off by entering NODATE.
ERRORS=*n*	Specifies the number of observations for which diagnostic information is printed in the DATA step. This information is printed whenever you attempt to read invalid data or perform an invalid calculation (e.g., division by zero). The usual default value of *n* is 20. It may be reset to any nonnegative number.
FIRSTOBS=*n*	Begin processing with observation *n* from all SAS datasets. The default value of *n* is 1. See also: OBS
FMTERR	Controls the SAS System's reaction to misspelled display formats. Use NOFMTERR if you want SAS to continue execution when you misspell a SAS format in a FORMAT statement. The default is to stop program execution.
LINESIZE=*n*	Use *n* print positions for procedure output. The default in interactive environments is usually 78. In batch, 132. *n* may range from 64 to 256.
MISSING=*'char'*	Specifies *char* as the value to display for missing numeric variables. The default is a period (.).
NUMBER	Numbers pages in the output listing. Turn off by entering NONUMBER.
OBS=*n*	Use up to and including observation *n* from SAS datasets. This option is useful when you want to test a program on a subset of your data. The default value of *n* is MAX (i.e., all available observations). See also: FIRSTOBS
PAGENO=*n*	Resets output page numbers. The next procedure's output will begin numbering from page *n*.
PAGES=*n*	Restricts output listings to *n* pages. This option is useful to prevent runaway jobs. The default value of *n* is MAX (i.e., no restriction). This option is not available in all environments.
PAGESIZE=*n*	Use *n* lines of output per page. The default in interactive environments is sufficient to fill the screen (that is, a "page" and a "screen" are identical). In batch, the default is usually about 66. *n* may range from 20 to 200.

DATA Step Programming Revisited

The DATA step is the SAS System's most versatile and powerful tool. We have touched on some of its capabilities in earlier chapters. Some additional features are discussed in this section. Note, however, that this book's length restrictions dictate that a *vast* number of details and features cannot be addressed. Appendix B has references to texts that provide more detail on this language within a language.

Grouping Related Statements: DO and END

So far, we have used relatively straightforward IF-THEN-ELSE structures. For each IF or ELSE condition, there was a single action, identified by a THEN clause. What if we needed to execute more than one statement if a condition were met? We could write code such as this:

```
IF sex = 'M' & inc > 50000 THEN group = 'PM';
IF sex = 'M' & inc > 50000 THEN OUTPUT;
```

This is not tedious, but is an invitation to coding errors because the condition must be repeated several times. SAS enables "bracketing" of any number of statements by using DO groups. The syntax is simple: Rather than enter an assignment, OUTPUT, or other statement after a condition, enter DO. Then enter one or more statements that should be executed only if the condition preceding the DO is met. End the DO group with, not surprisingly, an END statement. The previous example may be more cleanly written as

```
IF sex = 'M' & inc > 50000 THEN DO;
   group = 'PM';
   OUTPUT;
   END;
```

The only restriction on the use of DO groups is that the DATA step must contain an equal number of DO statements and END statements. Otherwise, SAS will issue an error message about "unclosed DO blocks" and generate an error. Note that DO groups may be contained within other DO groups. These are known as *nested* groups.

Streamlining Calculations: Functions

The DATA step is a powerful tool, one you can use to write virtually any calculation. In many instances, you can save yourself many lines of code by using the computational shortcuts known as *functions*. These are routines that come with the SAS System; they perform tasks such as univariate statistics, rounding, logarithms, and random number generation. This section identifies features of function use and then shows how to use some of the more frequently used functions.

Common Features With few exceptions, a function has two components. The first is the *name* of the function. It suggests what the function does: ROUND, SUM, and LOG, for example, perform rounding, addition, and logarithmic transformations. The second component is one or more *parameters* or *arguments* for the function to manipulate. Continuing with the functions noted above, the arguments to ROUND indicate the variable to be rounded and, optionally, the rounding unit (whole numbers, tenths, hundredths). SUM's arguments are two or more variables and/or constants to be added. LOG requires a single argument: the number for which a natural logarithm is needed.

There are a few items to remember when working with functions. First, the order of the parameters is important. ROUND, for instance, requires first the variable to be rounded, then the rounding unit. Incorrectly ordered parameters *may* create an error. If not, they will probably generate peculiar results. The second point to remember is the impact of missing values on function results. Many functions will use only nonmissing values. Compare the following assignment statements:

```
total = score_a + score_b + score_c;
total = sum(score_a, score_b, score_c);
```

In the first statement, a missing value for `score_a`, `score_b`, or `score_c` will set `total` to missing. When the `sum` function is used in the second statement, however, `total` will represent the sum of all nonmissing arguments.

Frequently Used Functions Exhibit 11.3 gives a summary of some of the more frequently used functions. They are listed alphabetically rather than grouped by function. Unless otherwise noted, all arguments are numeric variables or constants. Notice that when examples are provided, they use numeric constants for arguments. This is done to conserve presentation space. In most real-world applications, each argument would be all variables or a combination of variables and constants.

Exhibit 11.3

Frequently Used Functions

Function	Arguments	Comments
CEIL	*var*	Round *var* to the next largest integer. *Examples:* CEIL(8.3) is 9; CEIL(−2.3) is −2.
FLOOR	*var*	Round *var* to the next smallest integer. *Examples:* FLOOR(8.3) is 8; CEIL(−2.3) is −3.
LOG	*var*	Take the Naperian, or "natural," logarithm of *var*.
LOG10	*var*	Take the common, base-10, logarithm of *var*.
MEAN	*var1, var2, ...*	Computes the mean of a list of arguments. See additional comments in description of MAX, above. *Example:* MEAN(., ., 0, 2) is 1.

Exhibit 11.3

Frequently Used Functions (continued)

Function	Arguments	Comments
MAX	*var1, var2, ...*	Computes the maximum of a list of arguments. Any number of arguments may be specified. Argument order does not matter. Separate arguments with a comma. *Example:* MAX(., 2, 3, -2) is 3.
MIN	*var1, var2, ...*	Computes the minimum of a list of arguments. See additional comments in description of MAX, above. *Example:* MIN(., 2, 3, -2) is -2.
N	*var1, var2, ...*	Returns the number of arguments with nonmissing values. See additional comments in description of MAX, above. *Example:* N(., 2, 3, -2) is 3.
NMISS	*var1, var2, ...*	Returns the number of arguments with missing values. See additional comments in description of MAX, above. *Example:* NMISS(., 2, 3, 2) is 1.
RANNOR	*seed*	Generates a normally distributed (mean of 0, standard deviation of 1) random number. *seed* is optional. If entered, it must be 0 or a five- or seven-digit odd number.
RANUNI	*seed*	Generates a random number uniformly distributed between 0 and 1. *seed* follows the same rules as in RANNOR, above.
ROUND	*var, unit*	Rounds *var* to the nearest rounding *unit. unit* is optional. If omitted, it is set to 1. *Examples:* ROUND(1.5) is 2; ROUND(1.5, .5) is 1.5; ROUND(1.5, 2) is 2.
SQRT	*var*	Returns the square root of *var*.
STD	*var1, var2, ...*	Returns the standard deviation of arguments with nonmissing values. Requires at least two arguments. See additional comments in description of MAX, above.
SUM	*var1, var2, ...*	Returns the sum of arguments with nonmissing values. See additional comments in description of MAX, above.
VAR	*var1, var2, ...*	Returns the variance of arguments with nonmissing values. Requires at least two arguments. See additional comments in description of MAX, above.

Working with Dates

Some types of data represent the date on which an event occurred. You might want a variable or a constant to represent a survey respondent's date of birth, an employee's hire date, or a reference date such as the beginning of the current year. You might also want to manipulate these values, calculating the duration between events or answering questions such as these: How old is a person at a given date? How long has an employee been working at a given pay grade? Finally, you might want to extract information from a date. On which day of the week did an event occur? What was the most frequently occurring month for an event?

The SAS System provides a rich and powerful set of tools for reading, manipulating, and displaying date-oriented values. This section presents a brief overview of some of the more frequently used tools.

Background

The temptation is to create three different variables for each event — one for month, one for day, and one for year. Strictly speaking, this isn't wrong, but it does create more work than necessary for programmers of any level of sophistication. Just keep in mind a few basic things about the way SAS handles dates. First, the month, day, and year are stored in a single numeric variable. Second, the number represents the number of days before or after January 1, 1960. Events prior to that date are stored as negative numbers. Later dates are positive numbers. The "messy" aspects of dates (leap years, leap centuries, crossing year boundaries, and the like) are handled automatically by SAS.

Assigning Date Values

You might ask, "Why is this such a good thing?" After all, having to calculate the number of days since 1960 is not a trivial task. There are two pain-free ways to assign date values in the DATA step. The first is using a date constant. You can enter the date in a readable fashion, as follows:

```
'ddmmmyy'd
```

Here, `dd` is the day number (e.g., 3 or 23), `mmm` is the month name (e.g., jan or mar), and `yy` is the year (e.g., 95). The `d` tells SAS not to confuse the date constant with a character constant. The two constants below are *not* equivalent. The first is a numeric value (the number of days between February 12, 1994 and January 1, 1960). The second is a character value that looks like a date, but has none of the SAS System's date-handling features available to it.

```
'12feb94'd
'12feb94'
```

The second way to assign values to date-oriented data is to use special input formats, or informats on the INPUT statement. Chapter 3 discussed other numeric input formats. Exhibit 11.4 presents formats designed especially for dates. They permit you to read dates the way you would ordinarily write them. The dirty work — conversion of the date into the number of days since 1960 — is handled by SAS. Note the use of the separators in the examples. They help make the data more readable by separating the pieces of the date. You can use a slash (/), a hyphen (-), or a blank as a separator.

Exhibit 11.4

Date Input Formats

Format	Date Represented[1]	Width (min/max)	Examples Raw Data	Format
date	day/name/year	7/32	12feb94	date7.
ddmmyy	day/number/year	6/32	151095 15/10/95	ddmmyy6. ddmmyy8.
mmddyy	number/day/year	6/32	101595 10/15/95	mmddyy6. mmddyy8.
monyy	name/year	5/32	feb96 feb 96	monyy5. monyy6.
yymmdd	year/number/day	6/32	95-10-15	yymmdd8.

1. name = month name (e.g., March), number = month number (e.g., 3)

Manipulating Date Values

Some date manipulations may be carried out using simple arithmetic operators. Here are two examples:

```
interval = dose2 - dose1;
age = ('30jun94'd - birth_dt) / 365.25;
```

The variable `interval` is the difference between the dates of two administrations of a drug, `dose2` and `dose1`. The missing-value rule that applies to other arithmetic operations applies here as well: If `dose2` or `dose1` is missing, the result, `interval`, will be set to missing. The second calculation takes the difference between the constant `'30jun94'd` and the variable `birth_dt`, dividing by 365.25 to yield a value in years (there are functions available to do this, but they are beyond the scope of this book). The variable `age` represents the number of years between date of birth and the target date, June 30, 1994.

Another approach to date manipulation is using functions. The first section of this chapter presented arithmetic, statistical, and other functions. SAS also supplies functions for creating date values and extracting information from them. Exhibit 11.5 summarizes some of the more frequently used functions. They are listed alphabetically.

Exhibit 11.5

Date-Handling Functions

Function	Arguments	Comments
DAY	*datevar*	Returns the day number from the argument, a date variable. *Example:* Assume TERMDT represents October 15, 1995. DAY(termdt) is 15.
MDY	*month, day, year*	Returns a numeric value representing the date specified by the arguments. Useful when you receive a dataset in which dates were stored as three separate variables. *Example:* MDY(st_mon, st_day, st_year)

Exhibit 11.5

Date-Handling Functions (continued)

Function	Arguments	Comments
MONTH	*datevar*	Returns the month number from the argument, a date variable. *Example:* Assume TERMDT represents October 15, 1995. MONTH(termdt) is 10.
QTR	*datevar*	Returns the quarter number from the argument, a date variable. *Example:* Assume TERMDT represents October 15, 1995. QTR(termdt) is 4.
TODAY	none	Returns the current date. Note that parentheses are needed to prevent confusion between the TODAY *function* and a *variable* named TODAY. DATE may be used instead of TODAY. *Example:* TODAY().
WEEKDAY	*datevar*	Returns the day of the week from the argument, a date variable. Sunday equals 1; Saturday equals 7.
YEAR	*datevar*	Returns the year number from the argument, a date variable. *Example:* Assume TERMDT represents October 15, 1995. YEAR(termdt) is 15.

Displaying Date Values

So far, so good. You are able to read date values and manipulate them, but what about displaying them? It does little good to read or calculate a date value known to SAS as 12783 when in fact you would like it displayed as 31DEC94. As you might suspect, there are SAS formats that make the display more readable. Exhibit 11.6 presents display formats that resemble the input formats of Exhibit 11.4. As you can see in the second DATE format example, an insufficient field width results in loss of data. Be aware that SAS does *not* print a warning or error message when this situation arises.

Exhibit 11.6

Date Formats

Format	Date Represented[1]	Width (min/max)	December 31, 1994 Format	Result
DATE	day/name/year	5/9	date7. date5.	31dec94 31dec
DDMMYY	day/number/year	2/8	ddmmyy6. ddmmyy8.	311294 31/12/94
MMDDYY	number/day/year	2/8	mmddyy6. mmddyy8.	123194 12/31/94
MONYY	name/year	5/7	monyy5. monyy7.	dec94 dec1994
YYMMDD	year/number/day	2/8	yymmdd6. yymmdd8.	941231 94/12/31

[1] name = month name (e.g., March), number = month number (e.g., 3)

There are several other date display formats. These do not have direct parallels with the input formats. They display dates in more extended forms. Bear in mind that as with the formats described above, SAS will abbreviate or eliminate information to fit your column specification.

The additional formats are displayed in Exhibit 11.7. In the Items Displayed column, "wk" indicates day of the week (e.g., Sunday), "m" indicates month, "d" indicates day, and "y" indicates year.

Exhibit 11.7

Other Date Formats

Format	Item(s) Displayed[1]	Width min/max	December 31, 1994 Width	Value Displayed
WEEKDATE	wk, m d, y	3/37	15	Sat, Dec 31, 94
			20	Sat, Dec 31, 1994
			30	Saturday, December 31, 1994
WEEKDATX	wk, m d y	3/37	15	Sat, 31 Dec 94
			20	Sat, 31 Dec 1994
			30	Saturday, 31 December 1994
WEEKDAY	day number	1/32	1	7
WORDDATE	m d, y	3/32	15	Dec 31, 1994
			20	December 31, 1994
			30	December 31, 1994
WORDDATX	m d y	3/23	15	31 Dec 1994
			20	31 December 1994
			30	31 December 1994

[1] wk = weekday (e.g., Sunday), m = month number, d = day number, y = year, day number = 1 (Sunday) through 7 (Saturday)

Recap

Syntax Summary

This chapter has introduced a means to assess and control the SAS environment; functions; DO groups; and techniques for handling date-oriented data. (See the syntax description in Chapter 1 for an explanation of the notation.)

```
PROC OPTIONS <SHORT> <OPTION=option_name>;
OPTIONS option_setting;
<IF-THEN-ELSE condition> DO;
   ... any number of statements ...
   END:
```

We also introduced SAS functions. Their syntax varies:

```
function()
function(OF argument <, argument>)
function(argument, <argument>)
```

Finally, we discussed the SAS System's capabilities for handing date-oriented data. New input and display formats were introduced, as well as the date constant:

```
'mmdddyy'd
```

Content Review

The SAS System gives you considerable control of its environment. Through the use of system options, you can modify numerous aesthetic and computational aspects of the way SAS works. Some of the more popular options have been described in this chapter. We also revisited DATA step programming, introducing DO and END statements (to group-related statements) and functions (to simplify calculations).

In the remainder of the chapter, we gave an overview of how to work with date-oriented data. The SAS System has very powerful tools to read, manipulate, and display this class of data.

PROCs That Create Datasets

Some research activities require that data be "massaged" prior to analysis. As we have seen in earlier chapters, the DATA step provides a powerful set of tools for such tasks. Sometimes, however, it is simpler to let PROCs do this preliminary work for you. The PROCs described in this chapter perform activities that are both vital to the research process and tedious if coded using a DATA step.

The MEANS procedure has options that enable you to compute univariate statistics for population subgroups. You can, for example, take the sum and mean of variables aggregating individual students into larger (classroom) groups. The STANDARD procedure standardizes data to a given mean and/or standard deviation. Finally, the RANK procedure classifies variables by percentiles or other ordinal values.

Each of these procedures differs somewhat from those described in earlier chapters. Most of those procedures simply read in data and analyzed it. In the context of this chapter, MEANS, STANDARD, and RANK read a dataset and the output is another SAS dataset. This chapter describes their use and presents examples demonstrating their use in typical research applications.

Aggregating by Groups: The MEANS Procedure

The MEANS procedure, as we have used it so far, is used for the calculation of univariate statistics, possibly by a classification, or stratifier, variable. A simple extension of the syntax gives you the ability to save the results in a dataset for further analysis. First, let's look at an example using most of the MEANS procedure's defaults. Exhibit 12.1 contains a partial listing of the input dataset, the MEANS code, and a listing of the output dataset.

Exhibit 12.1

Using MEANS to Create an Output Dataset

```
First 15 observations from fish dataset

          SPECIES     WEIGHT     LEN1     LEN2     LEN3

             1         13.4       242     23.2     25.4
             1         13.8       290     24.0     26.3
             1         15.1       340     23.9     26.5
             1         13.3       363     26.3     29.0
             1         15.1       430     26.5     29.0
             1         14.2       450     26.8     29.7
             1         15.3       500     26.8     29.7
             1         13.4       390     27.6     30.0
             1         13.8       450     27.6     30.0
             1         13.7       500     28.5     30.7
             1         14.1       475     28.4     31.0
             1         13.3       500     28.7     31.0
             1         12.0       500     29.1     31.5
             1         13.6         .     29.5     32.0
             1         13.9       600     29.4     32.0
```

Program

```
PROC MEANS NOPRINT NWAY DATA=fish;
CLASS species;
VAR wgt l1 l2 l3;
OUTPUT OUT=summary SUM(wgt)=s_wgt
       MEAN(wgt l1 l2 l3) = m_wgt m_l1 m_l2 m_l3;
RUN;
```

Output

```
                FISH dataset summarized by species

OBS  SPECIES  _TYPE_  _FREQ_  S_WGT   M_WGT     M_L1      M_L2     M_L3

 1      1        1      35    494.6  14.1314  626.000  30.3057  33.1086
 2      2        1       6     95.4  15.9000  531.000  28.8000  31.3167
 3      3        1      20    292.1  14.6050  152.050  20.6450  22.2750
 4      4        1      11    154.9  14.0818  154.818  18.7273  20.3455
 5      5        1      14    143.1  10.2214   11.179  11.2571  11.9214
 6      6        1      17    177.4  10.4353  718.706  42.4765  45.4824
 7      7        1      56    887.0  15.8393  382.239  25.7357  27.8929
```

The PROC MEANS Statement

The PROC statement has the following syntax when MEANS is used to create output, or summary, datasets:

```
PROC MEANS DATA=dataset_name NWAY <NOPRINT>
     <MISSING>;
```

In this syntax,

- *dataset_name* is a one- or two-level name identifying the SAS dataset to be summarized.
- NWAY restricts the amount of detail in the output dataset.
- NOPRINT suppresses printed output. This is helpful when you run MEANS only to produce a summary dataset.
- MISSING allows missing values of the CLASS variables (described below) to be treated as valid categories. Normally, missing CLASS values are excluded from the output dataset.

Instructions for Aggregation: The CLASS Statement

MEANS needs to know how to aggregate the dataset. Use the CLASS statement to specify one or more variables according to whose levels the summary statistics will be computed. If you omit CLASS, the entire dataset will be summarized. The format is

```
CLASS varlist;
```

In the CLASS statement, *varlist* identifies one or more classification variables. The dataset does *not* need to be sorted by the variables in *varlist*.

There will be as many observations in the output dataset as there are combinations of values of the variables in *varlist*. Assume CLASS variable `race` has three levels and `region` has five levels. If the dataset contains all races in all regions, the output dataset will have 15 (3×5) observations. If a particular `race-region` combination does not exist, an observation *will not* be written to the output dataset. See "Examples," below, for an illustration of an output dataset with more than one CLASS variable.

Specifying Analysis Variables: The VAR Statement

The specification of the analysis variables is simple:

```
VAR varlist;
```

Here, *varlist* specifies numeric variables. To continue with the earlier example, if we wanted to analyze score and demographic variables, we might enter

```
CLASS race region;
VAR age income test1-test5;
```

Statistics for the variables `age`, `income`, and `test1` through `test5` will be computed for each combination of `race` and `region`.

Naming and Requesting Statistics: The OUTPUT Statement

The only remaining piece of this puzzle is naming the summary dataset and requesting the statistics. This is done with the OUTPUT statement:

```
OUTPUT OUT=output_dataset stat_request;
```

More than one OUTPUT statement can accompany a single execution of MEANS. In the statement,

- `output_dataset` is the name of the dataset that will contain the summary data. It can be either a one- or a two-level name. It can also be the same as the dataset specified by the DATA option in the PROC MEANS statement, although this practice is discouraged because the input data will be overwritten.
- `stat_request` identifies both the statistic you want and the name of the new variable in the output dataset. Some of the available statistics and their PROC MEANS keywords are presented in Exhibit 12.2.

Exhibit 12.2

OUTPUT Statement Statistics

Keyword	**For each level of CLASS variables ...**
n	Number of observations having *nonmissing* values
nmiss	Number of observations having *missing* values
min	Minimum value
max	Maximum value
range	Range of values
mean	Mean
std	Standard deviation
var	Variance
stderr	Standard error
cv	Coefficient of variation
skewness	Skewness
kurtosis	Kurtosis

Specify the variable names in the new dataset by using one or more of the formats shown in Exhibit 12.3. Examples follow the exhibit.

Exhibit 12.3

Specifying Variables in the MEANS Output Dataset

Request	Meaning
`stat=vars`	The variable list `vars` contains the statistic `stat` for each of the variables in the VAR statement. There is a one-to-one correspondence between the variables in the VAR statement and those in `vars`.
`stat(invars)=vars`	The variable list `vars` contains the statistic `stat` for the VAR statement variable(s) identified by `invars`.
`stat=`	The variables in the output dataset have the same name as the input (DATA=) dataset, only here they are the statistic `stat`, not the original value.
`stat(invars)=`	The VAR statement variable(s) identified by `invars` retain their name(s) in the output dataset, but contain the statistic `stat`, not the original value.

If you feel ill at ease with the prospect of coding the OUTPUT statement, a few examples should help ease your discomfort. All examples use the following VAR statement:

```
VAR age income test1-test5;
```

OUTPUT Example 1:

```
OUTPUT OUT=summdata MEAN=;
```

SUMMDATA contains the variables `age`, `income`, and `test1-test5`. The variables are the mean for each subgroup identified by the CLASS statement.

OUTPUT Example 2:

```
OUTPUT OUT=summdata MEAN(age income)=m_age m_inc
                    VAR(test1-test5)=vt1-vt5;
```

Here we calculate statistics more selectively than in Example 1. `m_age` and `m_inc` are the means of `age` and `income`. `vt1-vt5` are the variances of `test1` through `test5`.

OUTPUT Example 3:

```
OUTPUT OUT=summdata MEAN(age income)=;
```

Dataset `summdata` contains variables `age` and `income`. They are the means of `age` and `income` for each combination of CLASS statement variables.

OUTPUT Example 4:

```
OUTPUT OUT=summdata STD=std_inc std_age stdt1-stdt5;
```

SUMMDATA contains mislabeled variables. Compare the order of the VAR statement variables and the order of those in the OUTPUT statement. VAR begins with `age` and `income`, whereas the STD list begins with variable names implying `income` and `age`. The order is reversed, and SAS does not (indeed, should not!) flag this as an error. It's the programmer's responsibility to make sure the VAR and OUTPUT variables match up.

Examples

In Exhibit 12.4 we revisit the fishery data used at the beginning of this chapter. We first omit the CLASS statement. The output dataset has one observation. It contains summary statistics for the entire dataset.

Exhibit 12.4

Create MEANS Output Dataset Without a CLASS Statement.

Program

```
PROC MEANS NOPRINT NWAY DATA=fish;
VAR wgt l1 l2 l3;
OUTPUT OUT=summary MEAN(wgt l1 l2 l3) = m_wgt m_l1 m_l2 m_l3;
RUN;
```

Output

```
FISH dataset - no CLASS statement

   OBS    _TYPE_    _FREQ_    M_WGT      M_L1      M_L2      M_L3

    1       0        159     14.1164   398.696   26.2472   28.4157
```

Exhibit 12.5 uses multiple CLASS variables. Notice the order of observations in the output dataset: it is the same as if we used the SORT procedure with BY variables identical to those in the CLASS statement. That is, the output dataset `sum_fish` appears to have been sorted by the CLASS variables `species` and `sex`. This is the standard arrangement of MEANS's output datasets.

Exhibit 12.5

Multiple CLASS Statement Variables

Program

```
PROC MEANS NOPRINT NWAY DATA=fish;
CLASS species sex;
VAR wgt l1 l2 l3;
OUTPUT OUT=SUMMARY MEAN(wgt l1 l2 l3) = m_wgt m_l1 m_l2 m_l3;
RUN;
```

Exhibit 12.5

Multiple CLASS Statement Variables (continued)

Output

```
FISH dataset - Two CLASS Variables

OBS   SPECIES   SEX   _TYPE_   _FREQ_    M_WGT      M_L1       M_L2       M_L3

  1      1       0       3        3     14.8333   965.000    35.3000    38.8333
  2      1       1       3        6     14.1833   705.000    31.4333    34.2500
  3      2       0       3        1     16.6000   800.000    33.7000    36.4000
  4      3       0       3        8     14.9625   164.625    21.0125    22.6625
  5      4       0       3        4     14.3000   222.000    21.4500    23.3000
  6      4       1       3        3     13.5000    90.000    16.0333    17.4000
  7      5       0       3        9     10.3556    12.700    11.7111    12.4667
  8      5       1       3        5      9.9800     8.440    10.4400    10.9400
  9      6       0       3        5     10.4200   967.400    46.9400    50.2800
 10      6       1       3        1      9.7000   345.000    36.0000    38.5000
 11      7       0       3       25     16.3080   582.000    30.7240    33.1520
 12      7       1       3        2     17.0000   562.500    29.6000    32.2500
                                  ====
                                    72
```

Exhibit 12.6

Use the MISSING Option in the PROC Statement.

Program

```
PROC MEANS NOPRINT NWAY DATA=fish MISSING;
CLASS species sex;
VAR wgt l1 l2 l3;
OUTPUT OUT=SUMMARY SUM(wgt)=s_wgt
          MEAN(wgt l1 l2 l3) = m_wgt m_l1 m_l2 m_l3;
RUN;
```

Output

```
                 FISH dataset - Two CLASS Variables
         MISSING Option Increases Number of Observations Used

OBS   SPECIES   SEX   _TYPE_   _FREQ_    M_WGT      M_L1       M_L2       M_L3

  1      1       .       3       26     14.0385   571.692    29.4692    32.1846
  2      1       0       3        3     14.8333   965.000    35.3000    38.8333
  3      1       1       3        6     14.1833   705.000    31.4333    34.2500
  4      2       .       3        5     15.7600   477.200    27.8200    30.3000
  5      2       0       3        1     16.6000   800.000    33.7000    36.4000
  6      3       .       3       12     14.3667   143.667    20.4000    22.0167
  7      3       0       3        8     14.9625   164.625    21.0125    22.6625
  8      4       .       3        4     14.3000   136.250    18.0250    19.6000
  9      4       0       3        4     14.3000   222.000    21.4500    23.3000
 10      4       1       3        3     13.5000    90.000    16.0333    17.4000
 11      5       0       3        9     10.3556    12.700    11.7111    12.4667
 12      5       1       3        5      9.9800     8.440    10.4400    10.9400
 13      6       .       3       11     10.5091   639.636    41.0364    43.9364
 14      6       0       3        5     10.4200   967.400    46.9400    50.2800
 15      6       1       3        1      9.7000   345.000    36.0000    38.5000
 16      7       .       3       29     15.3552   197.600    21.1690    23.0586
 17      7       0       3       25     16.3080   582.000    30.7240    33.1520
 18      7       1       3        2     17.0000   562.500    29.6000    32.2500
                                  ====
                                   159
```

The example in Exhibit 12.6 adds the MISSING option to the code run in the preceding exhibit. The observation count increases dramatically once we let missing values be a valid level of the `sex` classification variable because there are many observations with a missing value for `sex`.

Standardization of Data: The STANDARD Procedure

Standardization of interval- and ratio-scale numeric variables is a common preliminary step in data analysis. The data are transformed so that they have a similar mean and/or standard deviation. A special case of this analysis is conversion to a mean of 0 and a standard deviation of 1, known as *z scores*. STANDARD performs these transformations and provides some flexibility when handling missing values. It creates a dataset containing all the variables in the input dataset; some or all of the variables will be transformed by options in the PROC statement.

The PROC STANDARD Statement

Most of the work in the procedure is done in the PROC statement:

```
PROC STANDARD DATA=in_data OUT=out_data
              <MEAN=newmean> <STD=newstd>;
```

In this statement,

- *`in_data`* and *`out_data`* identify the input and output SAS datasets. Either may be a one- or two-level dataset name.
- *`newmean`* specifies the mean that the analysis variables should have in the new dataset. If the MEAN option is not used, the mean of the output dataset variables is identical to their mean in the input dataset. For Z scores, specify MEAN=0.
- *`newstd`* specifies the new standard deviation that the analysis variables should have in the new dataset. If the STD option is not used, the standard deviation of the output dataset variables is identical to their standard deviation in the input dataset. For Z scores, specify STD=1.

Typically, at least one of *`newmean`* and *`newstd`* is entered. The PROC statement has some options that control the handling of printed output and missing values:

- REPLACE eliminates missing values in the output dataset. If an observation has a missing value in the input dataset, it will be written to the output dataset with the mean of the variable. If the MEAN option is used in addition to REPLACE, the missing values will be replaced by *`newmean`*.

- ❑ NOPRINT suppresses display of printed output. If specified, the mean, standard deviation, and number of observations used will not be displayed. NOPRINT is usually used when STANDARD is being run only to produce an output dataset.

Specifying Analysis Variables: The VAR Statement

The analysis variables are specified in a VAR statement:

```
VAR varlist;
```

`varlist` contains any number of numeric variables. Its format is exactly the same as those of the VAR statements shown in other procedures.

Examples

Exhibit 12.7 illustrates the impact of standardizing and replacement options. Variable `len1` in the `fish` dataset is transformed. Notice the change in the observation with the missing value near the end of the table.

Exhibit 12.7

Comparison of STANDARD Options

Input Dataset	MEAN=0	MEAN=0 STD=0	REPLACE MEAN=0 STD=0
242	-156.7	-0.4	-0.4
290	-108.7	-0.3	-0.3
340	-58.7	-0.2	-0.2
363	-35.7	-0.1	-0.1
430	31.3	0.1	0.1
450	51.3	0.1	0.1
500	101.3	0.3	0.3
390	-8.7	-0.0	-0.0
450	51.3	0.1	0.1
500	101.3	0.3	0.3
475	76.3	0.2	0.2
500	101.3	0.3	0.3
500	101.3	0.3	0.3
.	.	.	0.0
600	201.3	0.6	0.6
600	201.3	0.6	0.6
700	301.3	0.8	0.8
700	301.3	0.8	0.8

Rank-Ordering Data: The RANK Procedure

Ranking variables within a group, or across an entire dataset, is a common data management and reporting task. One method for doing this is to sort the dataset by the variable of interest, then read the sorted dataset, computing percentiles and ranks. This is tedious, error-prone, and wasteful of computer resources, especially when more than one variable is involved. As you might suspect, the RANK procedure provides a quick and reasonably straightforward way to determine the magnitude of a value relative to other values in a dataset.

Background

Before examining RANK syntax, it is helpful to stand back from the mechanics of the procedure and consider what you want to do with it. There are a few questions to ask:

- *What type of ranking?* The values may be divided into quartiles, deciles, percentiles, and so on, or they may be fractional or expressed as percentages. Use the GROUPS option in the PROC statement to specify your choice.
- *How to handle ties?* Tied ranks can be assigned the mean of the rank with multiple observations, the highest value of the tied ranks, or the lowest. Use the TIES option in the PROC statement to choose a tie-handling rule.
- *Keep the original values?* The rank values from the dataset can be kept in the output dataset *along with* the original values, or the ranks can *replace* the original values. Use the RANKS statement to preserve the old values.

The PROC RANK Statement

When you are considering the above questions, it is considerably easier to cut through the thicket of RANK syntax. Notice that there are no provisions for the handling of missing values. Variables with missing values are always excluded from rank calculations.

```
PROC RANK DATA=in_data <OUT=out_data>
          <TIES=MEAN|HIGH|LOW>
          <GROUPS=n_groups>
          <PERCENT> <FRACTION>;
```

In the above syntax,

- `in_data` and `out_data` identify the input and output SAS datasets. Either may be a one- or two-level dataset name.
- TIES identifies a method for handling tied ranks. Three options are available: MEAN (the default), HIGH, and LOW. MEAN assigns the sum of the tied ranks

divided by the number of observations tied at that value. HIGH assigns the highest of the tied ranks; LOW assigns the lowest.

- ❑ *n_groups* specifies the number of groups. Use this option to create categories such as quartiles (GROUPS=4), deciles (GROUPS=10), and percentiles (GROUPS=100). Note that group numbering begins with 0 rather than 1.
- ❑ Specify FRACTION to assign fractional rankings. For example, what if a variable had 20 nonmissing values in the dataset and an observation's value ranked 15th? The value in the output dataset would be 15/20, or .75. The PERCENT option closely resembles FRACTION, except that values are multiplied by 100 (15/20 would yield 75.0 instead of .75).

At least one of GROUPS, FRACTION, and PERCENT should be specified. The analysis variables and their meanings in the output dataset are addressed by VAR and RANK.

Specifying Input and Output Variables: The VAR and RANK Statements

The RANK procedure makes a number of assumptions about variables to analyze. You can override the default actions by using the VAR and RANK statements:

```
VAR analysis_vars;
RANK rank_vars;
```

In these statements,

- ❑ VAR is optional, but recommended. If it is omitted, all numeric variables in the dataset will be ranked using the instructions in the PROC statement. Otherwise, *analysis_vars* is a list of numeric variables in the dataset.
- ❑ RANK is optional. If it is omitted, the original values of *analysis_vars* will be replaced, in the output dataset, by their ranked values. If RANK is used, *rank_vars* assigns names to the rank variables. There is a one-to-one correspondence between the names in *analysis_vars* and *rank_vars*. See the comments in the PROC MEANS OUTPUT Example 4, earlier in this chapter. Here, as in MEANS, it is your responsibility to make sure the variable names accurately reflect their content!

Note that a BY statement may accompany the RANK procedure. If it is used, *in_data* must be sorted by the specified variables. The ranks will be calculated within each level of the BY group variables. Thus, you should take care not to interpret similar rank values as being ties; they may be similar ranks in *different groups*.

Examples

Exhibit 12.8 illustrates the impacts of different GROUP and TIES values. The effect of the VAR and RANK statements is dealt with later, in Exhibit 12.9.

Exhibit 12.8

Comparison of RANK Statement Options

	TIES=MEAN				TIES=HIGH	TIES=LOW
Original Value	GROUPS =1	GROUPS =5	GROUPS =100	PER-CENT	GROUPS =5	GROUPS =5
13.40	0.00	1.00	25.00	25.47	1.00	1.00
13.80	0.00	1.00	34.00	34.28	1.00	1.00
15.10	1.00	3.00	69.00	69.50	3.00	3.00
13.30	0.00	1.00	22.00	22.96	1.00	1.00
15.10	1.00	3.00	69.00	69.50	3.00	3.00
14.20	0.00	2.00	43.00	43.40	2.00	2.00
15.30	1.00	3.00	75.00	75.47	3.00	3.00
13.40	0.00	1.00	25.00	25.47	1.00	1.00
13.80	0.00	1.00	34.00	34.28	1.00	1.00
13.70	0.00	1.00	31.00	31.45	1.00	1.00
14.10	0.00	2.00	40.00	40.88	2.00	1.00
13.30	0.00	1.00	22.00	22.96	1.00	1.00
12.00	0.00	0.00	19.00	19.50	0.00	0.00
13.60	0.00	1.00	8.00	28.62	1.00	1.00
13.90	0.00	1.00	6.00	36.79	1.00	1.00
15.00	1.00	3.00	3.00	63.84	3.00	3.00
13.80	0.00	1.00	4.00	34.28	1.00	1.00
13.50	0.00	1.00	26.00	26.73	1.00	1.00
13.30	0.00	1.00	22.00	22.96	1.00	1.00
14.80	1.00	2.00	56.00	56.92	2.00	2.00
14.10	0.00	2.00	40.00	40.88	2.00	1.00

Exhibit 12.9 illustrates the impact of the VAR and RANKS statements. Using them enables you to see both the value of a variable and its rank. Omitting RANKS causes the original variable values to be replaced by their ranked values in the output dataset.

Exhibit 12.9

Impact of VAR and RANKS Statements

Program

```
PROC RANK DATA=book.fish GROUPS=10;
VAR len1 weight;      * Original len1 and weight values are replaced by
                        deciles ;
RUN;
```

Exhibit 12.9

Impact of VAR and RANKS Statements (continued)

Output

OBS	SPECIES	LEN1	LEN2	LEN3	HT_PCT	WID_PCT	WEIGHT	SEX	R_LEN1	R_WEIGHT
1	1	242	23.2	25.4	30.0	38.4	13.4	.	4	2
2	1	290	24.0	26.3	31.2	40.0	13.8	.	5	3
3	1	340	23.9	26.5	31.1	39.8	15.1	.	5	6
4	1	363	26.3	29.0	33.5	38.0	13.3	.	5	2
5	1	430	26.5	29.0	34.0	36.6	15.1	.	6	6
6	1	450	26.8	29.7	34.7	39.2	14.2	.	6	4
7	1	500	26.8	29.7	34.5	41.1	15.3	.	6	7
8	1	390	27.6	30.0	35.0	36.2	13.4	.	5	2
9	1	450	27.6	30.0	35.1	39.9	13.8	.	6	3
10	1	500	28.5	30.7	36.2	39.3	13.7	.	6	3
11	1	475	28.4	31.0	36.2	39.4	14.1	.	6	4
12	1	500	28.7	31.0	36.2	39.7	13.3	.	6	2
13	1	500	29.1	31.5	36.4	37.8	12.0	.	6	1
14	1	.	29.5	32.0	37.3	37.3	13.6	1	.	2
15	1	600	29.4	32.0	37.2	40.2	13.9	1	7	3
16	1	600	29.4	32.0	37.2	41.5	15.0	.	7	6
17	1	700	30.4	33.0	38.3	38.8	13.8	1	7	3
18	1	700	30.4	33.0	38.5	38.8	13.5	.	7	2
19	1	610	30.9	33.5	38.6	40.5	13.3	.	7	2
20	1	650	31.0	33.5	38.7	37.4	14.8	.	7	5

Recap

Syntax Summary

In this chapter, we have reviewed three procedures that create SAS datasets. Their syntax is summarized below. (See the syntax description in Chapter 1 for an explanation of the notation.)

```
PROC MEANS DATA=dataset_name <NOPRINT>
           <MISSING> NWAY;
CLASS summary_variable ...;
<VAR analysis_variable ...;>
OUTPUT OUT=output_dataset stat_request;

PROC STANDARD DATA=dataset_name
              OUT=output_dataset
              <MEAN=new_mean> <STD=new_std_dev>
              <REPLACE> <NOPRINT>;
<VAR analysis_variable ...;>

PROC RANK DATA=dataset_name
          <OUT=output_dataset>
```

```
             <TIES=MEAN|HIGH|LOW>
             <GROUPS=n_groups>
             <PERCENT|FRACTION>;
   <VAR analysis_variable ...;>
   <RANK rank_variable ...;>
```

Content Review

The SAS System has several procedures that simplify tasks such as rank-ordering, aggregation, and standardizing. This chapter has introduced two procedures, RANK and STANDARD, and revisited MEANS. Each has several options that control the contents of a SAS dataset created by the procedure. We described each option and compared their effects. Knowing what procedures and options are available can save a great deal of tedious and error-prone DATA step coding.

Recoding and Labeling with PROC FORMAT

Many analytical and report-writing situations require some form of data recoding. A value of 1 may need to be displayed as "Totally disagree," or a range of values may need to be combined into a single category, such as "low income." In Chapter 4, we reviewed some of the DATA step tools for performing these tasks. IF-THEN-ELSE statements were used to create new variables whose values were either descriptive, text values, or groupings of ranges. As we noted in Chapter 4, this technique is straightforward but cumbersome. It is also error-prone when creating large numbers of categories, since all or part of a range may be inadvertently assigned to more than one group.

A simple and powerful solution to the IF-THEN shortcomings is creating *custom formats* (also known as user-written formats). We have already seen formats used in Chapter 3. They displayed numeric and character values with commas, dollar signs, and other symbols that improve the legibility of the data. These formats are sometimes referred to as system formats, since they are part of the SAS System. User-written formats let you supplement the library of system formats. This chapter presents rules for their syntax and uses both in DATA steps and in PROCs.

Concepts

User-written formats carry the same benefits and usage caveats as system formats. These, and some additional usage notes, are listed below.

- *Data values are unchanged.* It's important to remember that the value of the variable being displayed with a format is unchanged. It is only the *display* that is affected by the format. This means that you can use different formats to display a variable. You could, for example, use a COMMA format in one procedure, then a user-written format for the same variable in a later procedure.
- *Usage scenarios vary.* User-written formats may be used in DATA steps and procedures. The format can be part of an assignment statement in a DATA step. It can also be used to create a temporary variable in an IF statement's condition. In procedures, the format may be part of a WHERE statement to filter observations. It may also be used to control the analytical groupings used by the procedure. Each of these usage scenarios is illustrated below.
- *Everything becomes character data.* Variables created by user-written formats are, by default, character variables even if the the format is used with a numeric variable. This subtle but important feature is demonstrated below.
- *Multiple uses.* A single format may be associated with many variables. If, for example, format `incgroup` groups incomes into several categories, the format could be assigned to variables `income89` through `income95`.
- *Not everything must be formatted.* If specifications for variable grouping are incomplete, the ungrouped values are left as is. This is helpful when you are trying to identify invalid data values. For example, suppose your custom format instructed SAS to display values of 1 as "Male" and 2 as "Female". If you used the format with PROC FREQ, the categories would be "Male", "Female", and any other values encountered in the dataset. Values other than 1 and 2 would be displayed, alerting you to problems with invalid codes in the dataset.

Syntax

The FORMAT procedure creates user-written formats. It has many options, some of which are beyond the scope of this chapter. Presented below are the most frequently used options for recoding and simple displays. Invoke the procedure as follows:

```
PROC FORMAT;
```

More than one format may be defined in a single use of the FORMAT procedure. Each format requires a VALUE statement:

```
VALUE fmtname
      range = "text"
      <range = "text" ...>;
```

This simple syntax is deceptively powerful. In the above,

- `fmtname` identifies the name of your format. If it will be applied to a character variable, `fmtname` must be preceded by a dollar sign ($). The name can contain letters or numbers, but must not begin or end in a number. The following names are valid: `$x, x, $incgrp, $advlx`. The following names are *not* valid: `much2long, $grp1990, 1stgrp`.

- `range` specifies one or more values, which will be treated as the string `text`. Character values must be enclosed in quotation marks. Implied ranges (e.g., `1-4`, or `'a'-'f'`) may be entered for both character and numeric values. Individual values and implied ranges in `range` must be separated by a comma. The boundaries of the range can be altered to avoid inclusion of one or both of the endpoints. This is demonstrated in Exhibit 13.1. Finally, the special values LOW, HIGH, and OTHER indicate the lowest nonmissing value, the highest value, and all other values not identified in earlier range definitions.

  Any number of `range` specifications may be entered. If more than one is entered, be sure not to have the same value in more than one range. If FORMAT detects duplication, it will stop the program with an error message. Some examples are briefly discussed below.

Exhibit 13.1

Range Examples

Range Specification	Meaning
1 – 20 = "Low"	Values from 1 through 20 are treated as the string "Low".
1 <– 20 = "Low"	Values from, but not including, 1 through 20 are treated as "Low".
1 –< 20 = "Low"	Values from 1 up to, but not including, 20 are treated as "Low".
1–20, 30–40 = "Grp A"	Values 1 through 20 and 30 through 40 are treated as "Grp A".
1, 4, 5–10 = "Pass 1"	Values 1, 4, and 5 through 10 are treated as "Pass 1".
"a" – "z" = "Low"	"a", "b", and so on through "z" are treated as "Low".
1–1000 = "G1" 1001–2000 = "G2" other = "?"	Values 1 through 2000 are grouped as either "G1" or "G2". Any other value is treated as a question mark ("?").
1–1000 = "G1" 1001–2000 = "G2"	Values 1 through 2000 are grouped as either "G1" or "G2". Any other value is left unformatted.
"ME", "VT" = "N. Eng" "NJ", "ME" = "N'east"	This creates an error because "ME" is in both range specifications.

- ❑ `text` is the character string that will represent the meaning of the corresponding *`range`*. It may be up to 200 characters long.

Using Formats for Display

The examples in this section use custom formats to enhance the display of data. Later sections will exercise the formats' power to recode and group values. Exhibit 13.2 uses several custom formats. `$st` combines several Pacific-oriented state codes into the strings "Pacific" and "S'east". Notice that since a residual, or "Other", category was not specified, values other than those specified in the ranges will display as is, unformatted. The numeric formats `pass` and `dep` create several groups. The breakpoints could be determined *a priori* by theory or from UNIVARIATE output containing percentiles of interest.

Exhibit 13.2

Using PROC FORMAT to Enhance Variable Display

Program

```
LIBNAME sasdata 'c:\book\data\sas';

PROC FORMAT;
VALUE $st "CA", "OR", "WA", "HI", "AK" = "Pacific"
          "FL", "GA", "NC", "SC"       = "S'east"   ;
VALUE pass LOW      -  1250000 = "Low"
           1250000 <- HIGH     = "High"    ;
VALUE dep        0  - 10000 = 'under 10k'
           10000 <- 50000 = '10k - 50k'
           50000 <- HIGH  = 'Over 50k'   ;
RUN;

PROC PRINT DATA=sasdata.airport(OBS=15);
ID airport;
VAR st perf_dep passnger;
FORMAT st $st. perf_dep dep. passnger pass.;
RUN;
```

Output

```
AIRPORT                 ST         PERF_DEP      PASSNGER

HARTSFIELD INTL         S'east     Over 50k        High
BALTO/WASH INTL         MD         Over 50k        High
LOGAN INTL              MA         Over 50k        High
DOUGLAS MUNI            S'east     Over 50k        High
MIDWAY                  IL         Over 50k        High
O'HARE INTL             IL         Over 50k        High
DALLAS/FT WORTH INTL    TX         Over 50k        High
LOVE FIELD              TX         10k - 50k       High
STAPLETON INTL          CO         Over 50k        High
DETROIT CITY            MI         under 10k       Low
WAYNE COUNTY            MI         Over 50k        High
WILLOW RUN              MI         under 10k       Low
HONOLULU INTL           Pacific    Over 50k        High
INTERCONTINENTAL        TX         Over 50k        High
HOBBY                   TX         Over 50k        High
```

Exhibit 13.3 modifies the `$st` format. The addition of the catchall OTHER range forces values not found in previous ranges to be displayed as "Other". Notice the change in appearance of variable `st` in the listing.

Exhibit 13.3

Using OTHER Range Specification for Display Purposes

Program

```
LIBNAME sasdata 'c:\book\data\sas';

PROC FORMAT;
VALUE $st "CA", "OR", "WA", "HI", "AK" = "Pacific"
          "FL", "GA", "NC", "SC"       = "S'east"
          OTHER                        = "Other"
      ;
VALUE pass LOW      -  1250000 = "Low"
          1250000 <- HIGH    = "High"
      ;
VALUE dep       0   - 10000 = 'under 10k'
           10000 <- 50000 = '10k - 50k'
           50000 <- HIGH  = 'Over 50k'
      ;
RUN;

PROC PRINT DATA=sasdata.airport(OBS=20);
ID airport;
VAR st perf_dep passnger;
FORMAT st $st. perf_dep dep. passnger pass.;
RUN;
```

Output

```
AIRPORT                    ST        PERF_DEP     PASSNGER

HARTSFIELD INTL            S'east    Over 50k       High
BALTO/WASH INTL            Other     Over 50k       High
LOGAN INTL                 Other     Over 50k       High
DOUGLAS MUNI               S'east    Over 50k       High
MIDWAY                     Other     Over 50k       High
O'HARE INTL                Other     Over 50k       High
DALLAS/FT WORTH INTL       Other     Over 50k       High
LOVE FIELD                 Other     10k - 50k      High
STAPLETON INTL             Other     Over 50k       High
DETROIT CITY               Other     under 10k      Low
WAYNE COUNTY               Other     Over 50k       High
WILLOW RUN                 Other     under 10k      Low
HONOLULU INTL              Pacific   Over 50k       High
INTERCONTINENTAL           Other     Over 50k       High
HOBBY                      Other     Over 50k       High
ELLINGTON FIELD            Other     under 10k      Low
MC CARRAN INTL             Other     Over 50k       High
HOLLYWOOD-BURBANK          Pacific   10k - 50k      High
LONG BEACH                 Pacific   10k - 50k      Low
LOS ANGELES INTL           Pacific   Over 50k       High
```

Unlike the earlier exhibits, which took several values and displayed them as just a few categories, Exhibit 13.4 uses the fishery data to demonstrate a simple one-to-one mapping of a numeric code to a text string. In this case, the values of the variables `species` and `sex` are enhanced. Note the syntax of PRINT's FORMAT statement: system formats and user-written formats are used together. Once a user-written format is defined, it is referenced *exactly* the same as a SAS System format.

Exhibit 13.4

Using FORMAT for One-to-One Mapping

Program

```
LIBNAME sasdata 'c:\book\data\sas';

PROC FORMAT;
VALUE species 1 = "Abramis brama"
              2 = "Leusiscus idus"
              3 = "Leusiscus rutilus"
              4 = "Abramis bjrnka"
              5 = "Osmerus eperlanus"
              6 = "Esox lucius"
              7 = "Perca fluviatilis"    ;
VALUE sex 1 = "Male"   2 = "Female"
          . = "Missing"   ;
RUN;

PROC PRINT DATA=sasdata.fish(OBS=15);
VAR len1-len3 sex;
FORMAT species species. sex sex. len1-len3 6.1;
RUN;
```

Output

```
SPECIES              LEN1       LEN2       LEN3     SEX

Abramis brama       242.0       23.2       25.4     Missing
Abramis brama       290.0       24.0       26.3     Missing
Abramis brama       340.0       23.9       26.5     Missing
Abramis brama       363.0       26.3       29.0     Missing
Abramis brama       430.0       26.5       29.0     Missing
Abramis brama       450.0       26.8       29.7     Missing
Abramis brama       500.0       26.8       29.7     Missing
Abramis brama       390.0       27.6       30.0     Missing
Abramis brama       450.0       27.6       30.0     Missing
Abramis brama       500.0       28.5       30.7     Missing
Abramis brama       475.0       28.4       31.0     Missing
Abramis brama       500.0       28.7       31.0     Missing
Abramis brama       500.0       29.1       31.5     Missing
Abramis brama           .       29.5       32.0     Male
Abramis brama       600.0       29.4       32.0     Male
```

Analytical Use of Formats

The use of custom formats extends beyond simply enhancing the readability of listings. There are also some powerful and simple analytical applications, none of which require any extensions to the syntax and usage described above. To understand this new application of formats, consider how a typical procedure, FREQ, works. When it is building a table for a variable, it treats each distinct level as a separate entry in the table. When the variable is formatted, the table entries are the *formatted* values. A table of salaries, if unformatted, could have potentially hundreds of entries. If a format grouped the salaries into "Low", "Medium", and "High", the table would have three entries (plus, of course, any salary values not captured by the range specifications). Note that the format could also be used simply to improve the meanings of individual levels of the analysis variable (the one-to-one mapping discussed earlier).

The situation is similar with other procedures. A format associated with a CLASS variable in MEANS could reduce the number of analytical subgroups. A format used with a CHART procedure's HBAR variable could collapse values into a single bar.

Exhibit 13.5 revisits the airport data. It compares the appearance of the variable `st` with and without use of a custom format.

Exhibit 13.5

Using a Custom Format Analytically to Collapse Variable Values

Program

```
LIBNAME sasdata 'c:\book\data\sas';

PROC FREQ DATA=sasdata.airport;
TABLES st;
RUN;
```

Output

ST	Frequency	Percent	Cumulative Frequency	Cumulative Percent
AK	2	1.5	2	1.5
AL	3	2.2	5	3.7
AR	1	0.7	6	4.4
AZ	2	1.5	8	5.9
CA	14	10.4	22	16.3
CO	2	1.5	24	17.8
CT	1	0.7	25	18.5
DC	2	1.5	27	20.0

Program

```
PROC FORMAT;
VALUE $st "CA", "OR", "WA", "HI", "AK" = "Pacific"
          "FL", "GA", "NC", "SC"       = "S'east"
          OTHER                        = "Other"
     ;
RUN;

PROC FREQ DATA=sasdata.airport;
TABLES st;
FORMAT st $st.;
RUN;
```

Output

ST	Frequency	Percent	Cumulative Frequency	Cumulative Percent
Pacific	24	17.8	24	17.8
Other	90	66.7	114	84.4
S'east	21	15.6	135	100.0

Exhibit 13.6 uses the fishery data to show how even a simple one-to-one mapping of a format can enhance the analytical process. The variable `species`, stored as 1 through 7, is displayed in the frequency distribution as the (presumably) more intelligible Latin name.

Exhibit 13.6

Custom Formats Performing One-to-One Mapping of Descriptive Text

Program

```
LIBNAME sasdata 'c:\book\data\sas';

PROC FORMAT;
VALUE species 1 = "Abramis brama"
              2 = "Leusiscus idus"
              3 = "Leusiscus rutilus"
              4 = "Abramis bjrnka"
              5 = "Osmerus eperlanus"
              6 = "Esox lucius"
              7 = "Perca fluviatilis";
RUN;

PROC FREQ DATA=sasdata.fish;
TABLES species;
FORMAT species species.;
RUN;
```

Output

```
Species code

                                                  Cumulative  Cumulative
         SPECIES    Frequency     Percent     Frequency     Percent
-----------------------------------------------------------------------
Abramis brama             35        22.0            35        22.0
Leusiscus idus             6         3.8            41        25.8
Leusiscus rutilu          20        12.6            61        38.4
Abramis bjrnka            11         6.9            72        45.3
Osmerus eperlanu          14         8.8            86        54.1
Esox lucius               17        10.7           103        64.8
Perca fluviatili          56        35.2           159       100.0
```

Using Formats in DATA Steps

User-written formats have a number of applications within the confines of the DATA step. They can be used to create new variables or can be part of conditions in IF statements. Both activities require use of the PUT function, as follows:

```
PUT(variable, fmtname.)
```

Here, `variable` is the name of the variable whose values you want to display; `fmtname` is the name of the format to use. The usage is similar to formatting a variable in a procedure: you need to specify the location of the values (`variable`) and the way you want them displayed (`fmtname`). The difference is in *where* the values are sent: in a PROC, the formatted value goes to the *output listing*.

In the DATA step, results are written to a *variable*. Look at the following code:

```
PROC FORMAT;
VALUE region 100-120 = "N'land"
             121-300 = "Auck."
[other range specifications omitted]
RUN;

DATA phase2;
SET in.master;
sale_reg = PUT(branch, region.);
```

Values of `branch` are formatted with `region` to create variable `sale_reg`. The new variable will be character. Its length will be equal to that of the longest text string assigned in the `value` statement defining `region`.

The PUT function can be used in IF and WHERE statement conditions. Notice that in this context you do not need to create a new variable. The result of the PUT function is held as a sort of "transient" variable. This use of PUT is shown in Exhibit 13.7.

Exhibit 13.7

PUT Function Used as Transient Variable

Use of PUT Function	Meaning
`where put(rate, rategrp.)="H";`	Use observations whose formatted value of `rate` is "H".
`where put(rate, rategrp.) in ("H","M");`	Use observations whose formatted value of `rate` is either "H" or "M".

Exhibit 13.7

PUT Function Used as Transient Variable (continued)

`if put(class, $class.) = "T" & failure <= .10 then do;`	Execute the DO group if the formatted value of `class` is "T" and `failure` is less than or equal to 10.
`if put(branch, region.)="Auck." then do;`	Execute the DO group only if the formatted value of `branch` equals "Auck.".

Recap

Syntax Summary

The FORMAT procedure offers a simple yet powerful set of commands to define custom formats. (See the syntax description in Chapter 1 for an explanation of the notation.)

```
PROC FORMAT;
VALUE <$>format_name
      range1 = "text1"
      <range2 = "text2">
      <other range-text strings as needed> ;
```

range may take the following forms:

```
"character value"
"character value", "character value"
"character starting value"-"character end value"
LOW - "character value"
"character value" - HIGH
number
number, number
number_start - number_end
LOW - number
number - HIGH
OTHER
```

Content Review

Formats, particularly user-written formats, are arguably the most powerful feature of the SAS programming language. In this chapter we have touched on the use of user-written formats to enhance the display of data, collapse multiple values of a variable into a single category, and perform other recoding activities. Effective use of the formats is broken into two phases: creating them in the FORMAT procedure and then using them in a DATA step or procedure. We discussed the syntax involved in both of these phases. Examples illustrated the flexibility of the formats. They may be used in a DATA step to create new variables; "on the fly" in a logical condition in a WHERE or IF-THEN-ELSE statement; or directly by a procedure.

Working with Character Data

Until now, we have had limited exposure to handling character data. Chapter 3 presented techniques for reading them, Chapter 4 demonstrated how to create character variables, and other chapters have used them analytically. Sometimes you need to work with this type of data a bit more rigorously. You may want to change a value's appearance or standardize it by converting it to uppercase, create a new variable by combining two or more character variables, or eliminate extraneous characters from a value. This chapter describes some of the underlying principles of working with character data and then illustrates SAS's tools for working with this type of data.

Principles

When working with character data, keep in mind two important characteristics: length and magnitude. This section discusses these features.

Length

The variable *length* refers to a variable's maximum number of characters and thus to the amount of disk storage it requires. In some cases, the amount is obvious: state postal codes will always need two characters. Industry- or company-specific codes may also have standard lengths. There may not be standards for other types of information, such as comment fields, names, and addresses. In either situation, you can specify the number of characters either in the INPUT statement or with a LENGTH statement (described in the following section).

Lengths are a bit less obvious when they are created by assignment statements. Look at the following example:

```
IF           condition1 THEN newvar = "Short";
   ELSE IF condition2 THEN newvar = "Longer";
```

When `condition2` is met, `newvar` will have the value “Longe”, not “Longer”. SAS sets the length of `newvar` by using the length of the first assignment statement’s string. In this case, the length of the first assignment’s value was 5, so `newvar` was created as a character variable of length 5. Values longer than 5 are truncated without a note or warning in the Log.

Character values created by assignment statements using functions can also have unexpected lengths. Whether a character variable is being read from raw data or created by simple assignment statements or through functions, it is always best to be explicit. Tell SAS how many characters the variable requires by using the LENGTH statement (described in the next section).

Magnitude

Like their numerical counterparts, character variables have a predetermined order. That is, one character variable can be greater than or less than another. For some characters this is obvious: “a” is less than “d” because it precedes “d” in the alphabet. Other character comparisons are a bit more obscure: Is “,” less than “a”?

Fortunately, people have taken the time to think about these issues and have established *character sets*, which determine the order and rank of variables. Unfortunately, there are two sets: EBCDIC (used mostly in IBM mainframe computers) and ASCII (used almost everywhere else). The reason you should care about such matters is that comparisons will test differently depending on the character set in use. In ASCII, `'a' > '0'` is true. In EBCDIC, it is false. If you will be working with character data at any but the simplest level, you should check with your instructor or with help-desk personnel for the exact order of characters in your computer system.

Specifying Length: The LENGTH Statement

The previous section discussed some of the perils and uncertainties of handling character data without explicitly specifying the variable’s length. The LENGTH statement addresses this need. Some variations of its syntax are shown below:

```
LENGTH var1 $x;
LENGTH var1 var2 $x;
LENGTH var1 $x var2 $y;
```

In the above, `var` specifies a character variable and `x` and `y` are variable lengths, or the numbers of columns or characters needed to hold each variable. Notice that a period does not follow the length specification ($12 is correct, $12. is not). Any number of LENGTH statements may be used in a single DATA step. Let's look at a few examples.

```
LENGTH first last $15;
```

Variables `first` and `last` are given a length of 15 characters.

```
LENGTH first last $15 init $1 stcode $2;
```

`first` and `last` have length 15, `init` has length 1, and `stcode` has length 2.

Character-Handling Operators

We have already seen that familiar comparison operators may be used to compare character values. Expressions such as both `st = "nc"` and `st > " "` are legitimate. There are a few extensions to SAS's syntax when you are dealing with character data. One enables you to combine two or more character variables into a single variable. The other restricts the effect of comparison operators. Both are presented below.

The Concatenation Operator

Situations sometimes arise in which you need to combine the contents of character variables and constants. You may want to combine first and last names, an ID code and a sequence number, and so on. Use the concatenation operator for this. It is entered as two vertical bars (||). Bear in mind that the variable being created is the sum of the lengths of the variables being concatenated (unless the new variable has been defined in a LENGTH statement). This means that if the sum exceeds the 200-character maximum, the 201st through last positions will be lost. No warning is printed in the Log when this occurs.

Also note that trailing blanks are kept in the new variable unless you explicitly remove them with the TRIM function (described later in this chapter). Review the following program fragments:

```
INPUT state $2.;
stname = state || ", USA";
```

The variable `stname` will be seven characters long: two characters from `state` and five from the constant. Contrast the building of `stname` to that of `fullname`, below:

```
INPUT first $ 1-15 last $ 16-30;
fullname = first || last;
```

`fullname` will be 30 characters long. It will probably look unattractive, since the trailing blanks will unnecessarily pad out the first name. If `first` and `last` were "John Q." and "Public", `fullname` would be:

```
John Q.        Public
```

See the discussion of the TRIM function later in this chapter for a technique to make `fullname` more presentable.

The Colon (:) Comparison Modifier

You may want to use comparison operators in order to use only the beginning portion of a character variable or constant. This arises in situations in which you need two-digit ratings beginning with "A", product names starting with "IBM", or three-digit credit scores of "B" or higher. Adding a colon (:) restricts the comparison to just the characters in the constant. Omitting it instructs SAS to pad out the rest of the variable or constant with blanks. Exhibit 14.1 shows the impact of the colon.

Exhibit 14.1

Using the Colon Modifier

Comparison Results				
Value of GR	**GR = "A"**	**GR =: "A"**	**GR > "A"**	**GR >: "A"**
A	True	True	False	False
A1	False	True	True	False

Character-Handling Functions

As with numeric data and their manipulation, the handling of character data is greatly simplified by using functions that are part of SAS. At first blush, it may not be obvious how or why you would want to manipulate character strings. A few examples may begin to give some idea of the range of applications and the tools available. An example at the end of this chapter demonstrates the "collaborative" nature of these character-handling tools. Typically, more than one function is used to obtain results. Before extracting part of a string, for example, you may need to know where the desired portion begins.

The functions discussed later in this section fall into the four groups presented below. For each group, we identify the related functions and present a quick sketch of the types of character handling performed by the group.

- *Variable information.* A variable should always be five columns long. If it is not, you do not want the observation used. *Function:* LENGTH.
- *Extract part of a string.* A variable contains a lab code in positions 5 through 7 of an identifier variable. *Function:* SUBSTR.
- *Locating strings within a string.* The variable `allareas` contains more than one area identifier, separated by commas. You want to test for the presence of a specific area. *Functions:* INDEX, INDEXC, INDEXW, VERIFY.
- *Changing string appearance.* A variable may have been entered with extraneous characters or in mixed case. You must standardize its appearance before analysis can begin. *Functions:* COMPRESS, COMPBBL, TRIM, UPCASE, LOWCASE, LEFT, RIGHT.

Obtaining Variable Information: LENGTH

The LENGTH *function* returns a numeric value indicating the length of the string. This may be different from the value specified in the LENGTH *statement.* The LENGTH *statement* establishes what is, in effect, the maximum number of columns used to store values. The LENGTH *function* indicates the rightmost nonblank character in the string. If the variable `status` were assigned a length of $5 in the LENGTH statement and an observation contained a `status` of "Pass", the LENGTH function would return a value of 4. The syntax of the function is

```
LENGTH(string)
```

where `string` is a character variable. In the following example, we test for a variable meeting a length requirement. If the requirement is not met, the observation is not written to the output dataset.

```
IF LENGTH(biocode) = 5 THEN OUTPUT;
```

Extracting Part of a String: SUBSTR

The SUBSTR function extracts all or, more commonly, part of a string. Its syntax is

```
SUBSTR(string, start, length)
```

Extraction begins in position `start` of *`string`* (a variable or constant). Extraction continues for *`length`* columns. If *`length`* is not specified, the remainder of *`string`* is extracted. Both *`start`* and *`length`*, if specified, must be numeric values greater than zero. Missing values for either parameter cause a note to be printed in the Log and force SUBSTR to return a missing value.

Suppose variable `pat_id` has length 13 and contains the clinic number in positions 10 through 13. Either of the following statements would correctly create `clinic`:

```
clinic = SUBSTR(pat_id, 10, 4);
clinic = SUBSTR(pat_id, 10);
```

Locating Strings Within a String: INDEX Variants and VERIFY

Our use of SUBSTR assumed that we knew where to look in a string. The "real world" is usually not so neat and tidy. Other situations arise in which you are interested in seeing if a string contains a particular substring. For example, if a variable contained place names, you might want to identify capes and bays. You would need to see if the substring "Cape" or "Bay" was located anywhere in the string. The INDEX and VERIFY functions and some of their variations address these needs.

Syntax All the functions require the same pieces of information: the name of the character variable to be examined and one or more substrings to be located.

```
INDEX(string, sub);
INDEXC(string, sub1 <, sub2>);
INDEXW(string, sub);
VERIFY(string, sub1 <, sub2>);
```

Here, *`string`* identifies the character variable to be examined; *`sub`*, *`sub1`*, and *`sub2`* are character variables or constants. The meaning of the value returned by the functions varies:

- *INDEX*. If *`sub`* is found in *`string`*, INDEX returns the column number where `sub` begins. Otherwise, it returns 0.
- *INDEXC*. If any of *`sub1`*, *`sub2`*, or other specified strings are located in *`string`*, INDEXC returns the column number where the substring begins. Otherwise, it returns 0.

- *INDEXW.* If *sub* is located in *string* as a word, INDEXW returns the column number where the substring begins. Otherwise, it returns 0. Consider this value of variable complnt:

```
had headache all day
```

  Now suppose we use the following statements in a DATA step:

```
acheloc1 = INDEX(complnt, "ache");
acheloc2 = INDEXW(complnt, "ache");
```

  The variable acheloc1 equals 9; acheloc2 equals 0, since "ache" was not a separate word in complnt.

- *VERIFY.* The INDEX family of functions looks for the *presence* of a substring; VERIFY looks for the string's *absence*. If any character in *sub1*, *sub2*, or other specified strings is not found in *string*, VERIFY returns the column location of the character. Otherwise, it returns 0. VERIFY is frequently used to locate invalid values in a string.

Examples Notice how many of these examples typically use more than one character-handling function to accomplish a task. The first example sets a flag to "T" if any of several strings are found. Notice that we do not need to create a new variable: the INDEXC function's results can be used "on the fly."

```
IF INDEXC(st,"VT","NH") > 0 THEN north = "T";
   else                          north = "F";
```

Group values of a test score:

```
IF          INDEXC(scr_pt1,"A","B") THEN grp = "H";
   ELSE IF INDEXC(scr_pt1,"C","D") THEN grp = "M";
   ELSE IF scr_pt1 = "F"           THEN grp = "L";
```

The variable pat_id has length 13; it contains the clinic number in positions 6 through 9 and the patient number in positions 10 through 13. Extract the clinic–patient number portion of the string for clinic "008A":

```
loc008a = INDEX(PAT_ID, 6, 4);
IF loc008a THEN clin_pat = SUBSTR(pat_id, 6);
```

Altering the Appearance of a String

Analytical and aesthetic considerations sometimes require that you alter the appearance of a string. You may, for example, need to convert a string to uppercase or lowercase to standardize its appearance and make logical comparisons more reliable. A more presentation-oriented requirement is removal of multiple embedded blanks from a string. SAS provides several functions to perform these and related tasks.

COMPRESS Use the COMPRESS function to remove characters from a string. Its syntax is

```
COMPRESS(string, <"chars">)
```

This tells SAS to remove all occurrences of *chars* from *string*. The string is rewritten as if *chars* had never been present.

COMPBBL A shortcoming of COMPRESS is that it is not aware of words within a string. If a blank is specified in *chars*, all blanks are removed, even those separating words. The COMPBBL function avoids this quandary by converting multiple blanks to a single blank. Its syntax is

```
COMPBBL(string)
```

Here, *string* is the name of the character constant or variable to be manipulated.

TRIM The earlier discussion of the concatenation operator identified the trailing-blank problem in concatenated strings. The TRIM function removes these blanks, thus making the concatenation of character variables more visually appealing. The syntax is

```
TRIM(string)
```

Here, *string* is the name of the character constant or variable to be manipulated. Let's revisit and then expand on the example presented earlier.

```
INPUT first $ 1-15 last $ 16-30;
fullnam1 = first || last;
fullnam2 = TRIM(first) || TRIM(last);
fullnam3 = TRIM(first) || " " || TRIM(last);


fullnam1 → John Q.        Public
fullnam1 → John Q.Public
fullnam1 → John Q. Public
```

Changing Case: UPCASE, LOWCASE The INDEX examples above relied on the values of `state` being in uppercase. A safer approach in many situations is to convert the string to uppercase or lowercase. This ensures that entries such as "me" and "ME" are treated as equal. Use the UPCASE and LOWCASE functions to perform this conversion:

```
UPCASE(string)
LOWCASE(string)
```

Here, `string` is the name of the character data to be changed. Nonalphabetic characters are unaffected by these functions. A more reliable way to rewrite the first INDEXC example is shown below. The variable `up_st` contains the standardized, uniform representation of the state code. This ensures that upper-, lower-, and mixed-case representations of `st` will receive a `north` value of "T".

```
up_st = UPCASE(st);
IF INDEXC(upst, "VT","NH") > 0 THEN nth = "T";
   ELSE                             nth = "F";
```

Recap

Syntax Summary

The following statements and operators were introduced in this chapter's discussion of character data. (See the syntax description in Chapter 1 for an explanation of the notation.)

```
LENGTH var ... $n_columns <var ... $n_columns>;
string1 || string2
character_operand =: character_operand
```

Content Review

In Chapter 13, we discussed a special treatment of numeric data that made handling date-oriented values easy. In this chapter, we have addressed special features of SAS's other data type, character data. We described an operator used expressly for character variables and constants. We also described and presented examples for character-handling functions. These enable you to determine how long a variable is, extract part of a character variable, locate information within a character variable, and alter the variable's appearance.

Putting It All Together

15

In the previous chapters, we showed you how to perform a variety of tasks using the SAS System. We covered topics such as reading and manipulating data, labeling variables, constructing formats, saving and retrieving SAS system files, and using various utility and statistical PROCs. In this chapter we will show you how to combine these tools into a series of SAS programs. We present three example analyses of data, which deal with some of the real-world problems you might encounter when analyzing data obtained from surveys or experiments. We wish to stress that these examples were constructed to show you how to employ elements of the SAS language to accomplish common analytical tasks. They are *not* intended to serve as models of statistical thoroughness.

Predicting Opinion About Taxes

Our first example uses data taken from the 1993 General Social Survey, conducted periodically by the National Opinion Research Center (Davis and Smith, 1993). Let's say we are interested in predicting respondents' opinions as to whether the taxes they paid were too high, about right, or too low. Information was gathered on age, level of education, family income, and political party identification of respondents. Discussion of analyses of data arising from more complex sampling designs is beyond the scope of this text. Thus, we'll assume that we can treat the data as if they came from a simple random sample. We will use SAS to

- ❑ Read the raw data for 1993 respondents from a large data file which contains data for respondents from surveys taken from 1972 through 1993

- ❑ Produce frequency distributions for the original survey variables so they can be checked against those given in the published codebook referred to above
- ❑ Create a permanent SAS dataset that recodes some of the original variables for use in later PROCs
- ❑ Create some output formats that will be used to group data values for some crosstabulations
- ❑ Obtain bivariate crosstabulations between the dependent variable and the independent variables
- ❑ Create a temporary SAS dataset that constructs a series of dummy variables for use in a logistic regression model
- ❑ Estimate a logistic regression model that predicts whether respondents feel they pay too much in taxes, versus too little or about the right amount, using the rest of the variables mentioned above

Exhibit 15.1 shows the SAS code used to extract the raw data. It also produces frequency distributions for the questions about age, education, income, political party identification, and opinion about taxes. The column locations for the variables were obtained from the published codebook for the survey. Note the use of the subsetting IF statement in the DATA step. It selects just the respondents for the 1993 survey. To conserve space, we show only part of the output from the PROC FREQ.

Exhibit 15.1

Reading Survey Data and Producing Frequency Distributions

Program

```
LIBNAME survey  'c:\gss'
***** Read Selected Variables from Combined GSS file. Layout *****;
***** of raw data found in GSS codebook from Roper Center     *****;
DATA survey.gss93;

  INFILE  'c:\gss72_93.all' ;

  INPUT  year 1-2    age  88-89    educ  94-95    income  147-148
         partyid  236    tax  338 ;

  LABEL  year      = "Survey Year"
         age       = "Respondent's Age"
         educ      = "Respondent's Level of Education"
         income    = "Family Income Level";
***  Select only cases from 1993 survey ***;
  IF year = 93  THEN  OUTPUT ;
run;

***** Check codes for original survey variables ***** ;
PROC FREQ  DATA=survey.gss93 ;
     TABLES  educ  age  income  partyid  tax ;
TITLE 'Original Codes of GSS93 Variables' ;
RUN;
```

Exhibit 15.1

Reading Survey Data and Producing Frequency Distributions (continued)

Output

Original Codes of GSS93 Variables

EDUC	Frequency	Percent	Cumulative Frequency	Cumulative Percent
0	3	0.2	3	0.2
2	4	0.2	7	0.4
4	7	0.4	14	0.9
5	7	0.4	21	1.3
6	20	1.2	41	2.6

Some Output Omitted

EDUC	Frequency	Percent	Cumulative Frequency	Cumulative Percent
18	73	4.5	1550	96.5
19	25	1.6	1575	98.1
20	27	1.7	1602	99.8
98	4	0.2	1606	100.0

Some Output Omitted

Original Codes of GSS93 Variables

AGE	Frequency	Percent	Cumulative Frequency	Cumulative Percent
18	5	0.3	5	0.3
19	18	1.1	23	1.4
20	20	1.2	43	2.7
21	24	1.5	67	4.2
22	16	1.0	83	5.2

Some Output Omitted

AGE	Frequency	Percent	Cumulative Frequency	Cumulative Percent
86	5	0.3	1585	98.7
87	6	0.4	1591	99.1
88	3	0.2	1594	99.3
89	7	0.4	1601	99.7
99	5	0.3	1606	100.0

Cumulative	Cumulative			
INCOME	Frequency	Percent	Frequency	Percent
1	17	1.1	17	1.1
2	20	1.2	37	2.3
3	25	1.6	62	3.9
4	34	2.1	96	6.0
5	34	2.1	130	8.1
6	24	1.5	154	9.6

Exhibit 15.1

Reading Survey Data and Producing Frequency Distributions (continued)

Some Output Omitted

```
  20          108          6.7          1307          81.4
  21          160         10.0          1467          91.3
  22           71          4.4          1538          95.8
  98           57          3.5          1595          99.3
  99           11          0.7          1606         100.0

                                    Cumulative    Cumulative
PARTYID    Frequency     Percent     Frequency      Percent
------------------------------------------------------------
      0          227        14.1           227        14.1
      1          321        20.0           548        34.1
      2          190        11.8           738        46.0
      3          205        12.8           943        58.7
      4          158         9.8          1101        68.6
      5          299        18.6          1400        87.2
      6          180        11.2          1580        98.4
      7           17         1.1          1597        99.4
      9            9         0.6          1606       100.0

                                Cumulative    Cumulative
TAX    Frequency     Percent     Frequency      Percent
--------------------------------------------------------
  1          582        54.1           582        54.1
  2          436        40.6          1018        94.7
  3           13         1.2          1031        95.9
  8           29         2.7          1060        98.6
  9           15         1.4          1075       100.0

Frequency Missing = 531
```

In Exhibit 15.2, we show the code we used to create character variables for tax opinion and party identification. We also use PROC FORMAT to create formats for grouping values of the survey variables. We use the new character variables (`ntax`, `npartyid`) and the formats in a PROC FREQ to obtain well-labeled, bivariate crosstabulations and chi-square statistics for the crosstabs.

Note that at the end of the DATA step we substitute the SAS missing value (.) for the numeric codes used to represent missing values for the variables `age`, `educ`, and `income`. We found the numeric codes in the documentation for the survey data. Cleaning up of numeric codes used for missing values is often necessary when using data collected by academic or commercial survey research organizations.

We display the Log and FREQ output. Note, in the Log, that SAS reports the creation of each format in PROC FORMAT. Also note in line 88 that even though we misspelled the TITLE command associated with the PROC FREQ, SAS managed to figure out what we meant. Though you can't always count on SAS to resolve command misspellings, it is often able to resolve minor misspelling problems.

SAS produces warning notes about about the validity of the chi-square statistics for the tables presented in the PROC FREQ output. This was prompted by the very small number of respondents who felt that their taxes were too low. Notice the other warning. This was caused by the high number of missing data values for the original tax opinion variable (531). Clearly any interpretation of these tables needs to take into account the factors in the survey design that resulted in this large number of missing values.

Exhibit 15.2

Recoding Variable Values and Producing Crosstabulations

Log

```
37     ***** Build SAS dataset with recoded party and tax variables ***** ;
38     DATA survey.gss93rec;
39       SET survey.gss93 ;
40
41       *** Create a new character variable for tax ***;
42       IF      tax = 2 THEN ntax = 'About Right'; /* longest first */
43       ELSE IF tax = 1 THEN ntax = 'Too High';
44       ELSE IF tax = 3 THEN ntax = 'Too Low';
45       ELSE IF tax > 3 THEN ntax = ' '       /* Assign missing value */
46
47       *** Collapse party ID variable into 3 category character var ***;
48       IF       partyid = 0  OR  partyid = 1 THEN npartyid= Democrat    ';
49       ELSE IF  2 <=  partyid  <= 4           THEN npartyid='Independent';
50       ELSE IF  partyid = 5  OR  partyid = 6 THEN npartyid='Republican ';
51       ELSE npartyid = ' ' ;                 /* Assign missing value */
52
53       LABEL  ntax     = 'Opinion About Amount of Tax Paid'
54              npartyid = 'Respondent''s Political Party Leaning' ;
55
56       *** Assign missing values for age, education, and income *** ;
57
58       IF  age     = 99  THEN  age    = . ;
59       IF  educ   >= 98  THEN  educ   = . ;
60       IF  income >= 22  THEN  income = . ;
61
62     RUN;
NOTE: The data set SURVEY.GSS93REC has 1606 observations and 22 variables.
NOTE: The DATA statement used 8.78 seconds.
63

64     ***** Build formats for grouping age, education, and income variables
65           Note that no output format is assigned for missing values. ***;
66     PROC FORMAT  LIBRARY=LIBRARY;
67
```

Exhibit 15.2

Recoding Variable Values and Producing Crosstabulations (continued)

```
68        VALUE   agefmt LOW - 29   = 'Under 30'
69                        30 - 49   = '30 - 49'
70                        50 - 69   = '50 - 69'
71                        70 - HIGH = '70 & Over' ;
NOTE: Format AGEFMT has been written to LIBRARY.FORMATS.

72
73        VALUE   educfmt 00 - 11 = 'Less than H.S.'
74                             12 = 'H.S. Graduate'
75                        13 - 16 = '1 - 4 yrs. College'
76                        17 - 20 = 'Some Post Grad.' ;
NOTE: Format EDUCFMT has been written to LIBRARY.FORMATS.

77
78        VALUE   incfmt LOW - 10 =  'Under 20K'
79                        11 - 19 =  '20K to <60K'
80                        20 - 21 =  '60K & Over' ;
NOTE: Format INCFMT has been written to LIBRARY.FORMATS.

81
82   RUN;

NOTE: The PROCEDURE FORMAT used 3.45 seconds.

83
84   ***** Create bivariate crosstabs and obtain chi-square stats *****;
85   PROC FREQ  DATA=survey.gss93rec   ;
86   TABLES  (age educ income npartyid) * ntax / nocol nopercent chisq ;
87   FORMAT  age agefmt.  educ educfmt.  income incfmt. ;
88   TTILE 'Bivariate Relationships for Recoded Variables' ;
     -----
     14
89   RUN;

WARNING 14-169: Assuming the symbol TITLE was misspelled as TTILE.
WARNING: The formatted values of one or more variables are truncated to 16
         characters.
NOTE: For table location in print file, see
      page 4 for AGE*NTAX
      page 4 for EDUC*NTAX
      page 5 for INCOME*NTAX
      page 6 for NPARTYID*NTAX
NOTE: The PROCEDURE FREQ used 5.49 seconds.
```

Exhibit 15.2

Recoding Variable Values and Producing Crosstabulations (continued)

Output Begins

```
Bivariate Relationships for Recoded Variables

TABLE OF AGE BY NTAX

AGE        NTAX(Opinion About Amount of Tax Paid)

Frequency |
Row Pct   |Too High|About Ri|Too Low |
          |        |ght     |        |  Total
----------+--------+--------+--------+
Under 30  |    110 |     89 |      2 |    201
          |  54.73 |  44.28 |   1.00 |
----------+--------+--------+--------+
30 - 49   |    285 |    166 |      7 |    458
          |  62.23 |  36.24 |   1.53 |
----------+--------+--------+--------+
50 - 69   |    129 |    102 |      4 |    235
          |  54.89 |  43.40 |   1.70 |
----------+--------+--------+--------+
70 & Over |     57 |     76 |      0 |    133
          |  42.86 |  57.14 |   0.00 |
----------+--------+--------+--------+
Total          581      433       13     1027

Frequency Missing = 579
STATISTICS FOR TABLE OF AGE BY NTAX

Statistic                     DF       Value       Prob
------------------------------------------------------
Chi-Square                     6      20.835      0.002
Likelihood Ratio Chi-Square    6      22.278      0.001
Mantel-Haenszel Chi-Square     1       6.873      0.009
Phi Coefficient                        0.142
Contingency Coefficient                0.141
Cramer's V                             0.101

Effective Sample Size = 1027
Frequency Missing = 579
WARNING:  36% of the data are missing.
WARNING:  25% of the cells have expected counts less
           than 5. Chi-Square may not be a valid test.
```

Exhibit 15.2

Recoding Variable Values and Producing Crosstabulations (continued)

```
TABLE OF EDUC BY NTAX

EDUC              NTAX(Opinion About Amount of Tax Paid)

Frequency        |
Row Pct          |Too High|About Ri|Too Low |
                 |        |ght     |        |  Total
-----------------+--------+--------+--------+
Less than H.S.   |    110 |     84 |      0 |    194
                 |  56.70 |  43.30 |   0.00 |
-----------------+--------+--------+--------+
H.S. Graduate    |    184 |    123 |      2 |    309
                 |  59.55 |  39.81 |   0.65 |
-----------------+--------+--------+--------+
1 - 4 yrs. Colle |    225 |    176 |      7 |    408
                 |  55.15 |  43.14 |   1.72 |
-----------------+--------+--------+--------+
Some Post Grad.  |     61 |     53 |      4 |    118
                 |  51.69 |  44.92 |   3.39 |
-----------------+--------+--------+--------+
Total                 580      436       13     1029

Frequency Missing = 577
```

New Page of Output

```
Bivariate Relationships for Recoded Variables

STATISTICS FOR TABLE OF EDUC BY NTAX

Statistic                     DF       Value        Prob
------------------------------------------------------
Chi-Square                     6      10.143       0.119
Likelihood Ratio Chi-Square    6      11.394       0.077
Mantel-Haenszel Chi-Square     1       0.877       0.349
Phi Coefficient                        0.099
Contingency Coefficient                0.099
Cramer's V                             0.070

Effective Sample Size = 1029
Frequency Missing = 577
WARNING:  36% of the data are missing.
WARNING:  25% of the cells have expected counts less
           than 5. Chi-Square may not be a valid test.
```

Exhibit 15.2

Recoding Variable Values and Producing Crosstabulations (continued)

```
TABLE OF INCOME BY NTAX
INCOME        NTAX(Opinion About Amount of Tax Paid)
Frequency    |
Row Pct      |Too High|About Ri|Too Low |
             |        |ght     |        |  Total
-------------+--------+--------+--------+
Under 20K    |    101 |    121 |      1 |    223
             |  45.29 |  54.26 |   0.45 |
-------------+--------+--------+--------+
20K to <60K  |    338 |    211 |      6 |    555
             |  60.90 |  38.02 |   1.08 |
-------------+--------+--------+--------+
60K & Over   |    103 |     72 |      5 |    180
             |  57.22 |  40.00 |   2.78 |
-------------+--------+--------+--------+
Total             542      404       12      958

Frequency Missing = 648

STATISTICS FOR TABLE OF INCOME BY NTAX

Statistic                       DF     Value       Prob
------------------------------------------------------
Chi-Square                       4    21.691      0.000
Likelihood Ratio Chi-Square      4    20.970      0.000
Mantel-Haenszel Chi-Square       1     4.844      0.028
Phi Coefficient                        0.150
Contingency Coefficient                0.149
Cramer's V                             0.106

Effective Sample Size = 958
Frequency Missing = 648
WARNING:  40% of the data are missing.
WARNING:  22% of the cells have expected counts less
          than 5. Chi-Square may not be a valid test.
```

New Page of Output

```
Bivariate Relationships for Recoded Variables

TABLE OF NPARTYID BY NTAX

NPARTYID(Respondent's Political Party Leaning)
             NTAX(Opinion About Amount of Tax Paid)
Frequency    |
Row Pct      |Too High|About Ri|Too Low |
             |        |ght     |        |  Total
-------------+--------+--------+--------+
Democrat     |    186 |    166 |      4 |    356
             |  52.25 |  46.63 |   1.12 |
-------------+--------+--------+--------+
Independent  |    205 |    140 |      6 |    351
             |  58.40 |  39.89 |   1.71 |
-------------+--------+--------+--------+
Republican   |    179 |    127 |      3 |    309
             |  57.93 |  41.10 |   0.97 |
-------------+--------+--------+--------+
Total             570      433       13     1016

Missing = 590
```

Exhibit 15.2

Recoding Variable Values and Producing Crosstabulations (continued)

```
STATISTICS FOR TABLE OF NPARTYID BY NTAX
Statistic                          DF      Value          Prob
-----------------------------------------------------------
Chi-Square                          4      4.389         0.356
Likelihood Ratio Chi-Square         4      4.355         0.360
Mantel-Haenszel Chi-Square          1      2.155         0.142
Phi Coefficient                            0.066
Contingency Coefficient                    0.066
Cramer's V                                 0.046
Effective Sample Size = 1016
Frequency Missing = 590
WARNING:  37% of the data are missing.
WARNING:  33% of the cells have expected counts less
          than 5. Chi-Square may not be a valid test.
```

Not surprisingly, there were not many respondents who thought that their taxes were too low. We decided to use a logistic regression equation to assess the influence of all the independent variables on whether respondents felt their taxes were too high versus about right or too low. Exhibit 15.3 shows the SAS Log for the code we used to construct dummy variables for the categorical variables. It employs PROC LOGISTIC to estimate a logistic regression equation. We used the KEEP dataset option for dataset TEMP to keep only the variables needed for the logistic regression. This speeds processing, especially when dealing with datasets with large numbers of observations and/or variables.

Exhibit 15.3

Estimating a Logistic Regression Equation for Predicting Opinion that Taxes Are Too High

Log

```
92    ** Create temporary vars for logistic regression model **;
93    DATA temp(KEEP=tax0_1 age educ midincom upincom democ repub);
94      SET   survey.gss93rec ;
95
96      *** Create 0,1 valued var for opinion on taxes ***;
97      IF        tax = 1 THEN   tax0_1 = 1 ;          /* Too high */
98      ELSE IF tax = 2 OR tax=3 THEN tax0_1 = 0; /* Too low, OK */
99
100     *** Create new 0,1 dummy variables for income   ***;
101     *** Compare middle and upper income to lower.   ***;
102
103     IF        1 <= income <= 10 THEN midincom = 0; /* Mid inc */
104     ELSE IF 11 <= income <= 19 THEN midincom = 1 ;
105     ELSE IF 20 <= income <= 21 THEN midincom = 0 ;
106
107
108     IF        1 <= income <= 10 THEN upincom = 0;   /* UpperInc */
109     ELSE IF 11 <= income <= 19 THEN upincom = 0 ;
110     ELSE IF 20 <= income <= 21 THEN upincom = 1 ;
111
112     *** Create new 0,1 dummy var for party identification ***;
113     *** Compare Democrats to Republicans and Independents ***;
114
115     IF        npartyid = 'Democrat' THEN       democ = 1;
116     ELSE IF   npartyid = 'Republican' OR
117               npartyid = 'Independent'   THEN   democ = 0 ;
```

Exhibit 15.3

Estimating a Logistic Regression Equation for Predicting Opinion That Taxes Are Too High (continued)

```
118
119
120    IF        npartyid = 'Republican' THEN   repub = 1; /* Republican
             */
121    ELSE IF  npartyid = 'Democrat'  OR
122             npartyid = 'Independent' THEN  repub = 0 ;
123  RUN;

NOTE: The data set WORK.TEMP has 1606 observations and 27 variables.
NOTE: The DATA statement used 7.25 seconds.

124
125
126  ***** Estimate logistic regression model using vars created above
127        Use DESCENDING option to predict "Too high" response. *****;
128  PROC LOGISTIC  DATA=temp  DESCENDING;
129       MODEL  tax0_1  = age  educ midincom upincom democ repub ;
130  TITLE1 "Logistic Regression Model Predicting" ;
131  TITLE2 "Respondent's Opinion That Taxes Are Too High" ;
132  RUN;

NOTE: The PROCEDURE LOGISTIC used 7.46 seconds.
```

New Page of Output

```
Logistic Regression Model Predicting
Respondent's Opinion That Taxes Are Too High

The LOGISTIC Procedure

Data Set: WORK.TEMP
Response Variable: TAX0_1
Response Levels: 2
Number of Observations: 942
Link Function: Logit

     Response Profile

Ordered
  Value  TAX0_1     Count

      1       1       529
      2       0       413

WARNING: 664 observation(s) were deleted due to missing values for the
         response or explanatory variables.

                     Criteria for Assessing Model Fit

                            Intercept
              Intercept        and
Criterion       Only       Covariates    Chi-Square for Covariates

AIC           1293.568      1268.416              .
SC            1298.416      1302.352              .
-2 LOG L      1291.568      1254.416        37.153 with 6 DF (p=0.0001)
Score             .             .           36.696 with 6 DF (p=0.0001)
```

Exhibit 15.3

Estimating a Logistic Regression Equation for Predicting Opinion That Taxes Are Too High (continued)

```
                 Analysis of Maximum Likelihood Estimates

                Parameter Standard    Wald       Pr >    Standardized     Odds
Variable   DF   Estimate   Error   Chi-Square Chi-Square   Estimate      Ratio
INTERCPT    1     1.6385   0.4200    15.2230    0.0001              .    5.148
AGE         1    -0.0126  0.00408     9.6189    0.0019      -0.119163    0.987
EDUC        1    -0.0958   0.0259    13.6641    0.0002      -0.156239    0.909
MIDINCOM    1     0.7135   0.1732    16.9709    0.0001       0.194149    2.041
UPINCOM     1     0.7154   0.2302     9.6592    0.0019       0.153150    2.045
DEMOC       1    -0.2502   0.1615     2.3988    0.1214      -0.066066    0.779
REPUB       1    -0.0143   0.1689     0.0071    0.9326      -0.003635    0.986

Association of Predicted Probabilities and Observed Responses

  Concordant = 61.1%          Somers' D = 0.228
  Discordant = 38.3%          Gamma     = 0.229
  Tied       =  0.6%          Tau-a     = 0.112
  (218477 pairs)              c         = 0.614
```

Predicting Employee Turnover

Our next example involves a regression analysis of data taken from a survey of benefits provided to employees of a sample of small businesses (McLaughlin, 1990). The survey also gathered data on employee turnover. We could use this information to see if the provision of greater benefits was related to lower employee turnover. In this example we will

- ❑ Use a DATA step to prepare an analysis dataset from another SAS dataset containing the original survey data. The DATA step will convert numeric missing values in the raw data to SAS missing values and create some dummy variables for the regression analysis.
- ❑ Examine and transform the distribution of the dependent variable in the raw data.
- ❑ Plot some relationships between indepdendent variables and the transformed dependent variable.
- ❑ Produce a matrix of correlations among the analysis variables.
- ❑ Estimate a multiple regression equation relating the transformed dependent variable to the independent variables.

Assume that the original, raw survey data were read by a SAS program and placed in a SAS dataset called BENEFITS. We begin the program in Exhibit 15.4 by using a DATA step to assign missing values to the original survey variables. If we did not do this, SAS could not distinguish between legitimate numeric values and those that were coded to represent missing values. Next, we recode the variables that contained responses to questions as to whether particular benefits were offered. These are converted into 0- or 1-valued dummy variables. We also adjust the value of `hlthpct` to 0 if a firm didn't offer health insurance. PROC UNIVARIATE checks the distribution of the variable `ftpleft`, the number of full-time employees who left a firm. You can see from the plotted output from PROC UNIVARIATE that this variable has a highly skewed distribution because many of the firms had no turnover.

Exhibit 15.4

First Analysis of Benefits Survey

Log

```
5     LIBNAME  data  'c:\sasbook\datasets'  ;
NOTE: Libref DATA was successfully assigned as follows:
      Engine:        V608
      Physical Name: C:\SASBOOK\DATASETS
6
7     ***** Prepare Data for Analysis *****;
8     DATA temp1 ;
9       SET data.benefits ;
10
11      LABEL  firmid   = 'Firm ID Number'
12             nworkers = 'Total Number of Workers'
13             ftphired = 'FT Perm. Workers Hired in Last Year'
14             ftpleft  = 'FT Perm. Workers Left in Last Year'
15             sicklv   = 'Offer Sick Leave'
16             disablty = 'Offer Disability Plan'
17             lifeins  = 'Offer Life Insurance'
18             retire   = 'Offer Retirement Plan'
19             healthin = 'Offer Health Insurance'
20             hlthpct  = 'Percent Health Ins. Paid by Firm'
21             gross    = 'Gross Receipts in Last Year' ;
22
23      *** Assign SAS Missing Values Based on Study Codebook ***;
24      IF ftphired = 99  THEN  ftphired = . ;
25      IF ftpleft  >= 98 THEN  ftpleft  = . ;
26      IF sicklv   >= 8  THEN  sicklv   = . ;
27      IF disablty >= 8  THEN  disablty = . ;
28      IF lifeins  >= 8  THEN  lifeins  = . ;
29      IF retire   >= 8  THEN  retire   = . ;
30      IF healthin >= 8  THEN  healthin = . ;
31      IF hlthpct >= 996 THEN  hlthpct  = . ;
32      IF gross    >= 8  THEN  gross    = . ;
33
34
35      *** Recode benefits to  0,1 dummy vars       ***
36      *** 1 & 2 = Offered to all or some employees ***
37      ***      3 = Not offered                     *** ;
38      IF       sicklv   = 1  OR sicklv   = 2 THEN sicklv   = 1 ;
39      ELSE IF  sicklv   = 3                  THEN sicklv   = 0 ;
40
41      IF       disablty = 1  OR disablty = 2 THEN disablty = 1 ;
42      ELSE IF  disablty = 3                  THEN disablty = 0 ;
43
44      IF       lifeins  = 1  OR lifeins  = 2 THEN lifeins  = 1 ;
45      ELSE IF  lifeins  = 3                  THEN lifeins  = 0 ;
46
47      IF       retire   = 1  OR  retire  = 2 THEN retire   = 1 ;
48      ELSE IF  retire   = 3                  THEN retire   = 0 ;
```

Exhibit 15.4

First Analysis of Benefits Survey (continued)

```
49
50
51      *** Percent of health insurance paid by firm ***
52      *** Percent is also 0 if firm does not offer ***
53      *** 1 = health ins. offered  2 = not offered *** ;
54      IF        healthin = 1  THEN  hlthpct  = hlthpct;
55      ELSE IF  healthin = 2  THEN  hlthpct  = 0;
56
57    RUN;

NOTE: The data set WORK.TEMP1 has 1364 observations and 12 variables.
NOTE: The DATA statement used 13.28 seconds.

58
59
60    *** Examine distribution of number of FT employees who left *** ;
61    PROC UNIVARIATE   PLOT  ;
62         VAR  ftpleft       ;
63
64    TITLE1 'Analysis of Influence of Employee Benefits on Turnover' ;
65    TITLE2 'Distribution of Number of FT Employees Who Left' ;
66    RUN;

NOTE: The PROCEDURE UNIVARIATE used 2.91 seconds.
```

Output Begins

```
Analysis of Influence of Employee Benefits on Turnover
Distribution of Number of FT Employees Who Left

Univariate Procedure

Variable=FTPLEFT        FT Perm. Workers Left in Last Year

                   Moments

N                1360  Sum Wgts       1360
Mean         1.480147  Sum            2013
Std Dev      4.132951  Variance   17.08128
Skewness      11.6971  Kurtosis   186.8201
USS             26193  CSS        23213.46
CV           279.2257  Std Mean    0.11207
T:Mean=0      13.2073  Pr>|T|       0.0001
Num ^= 0          696  Num > 0         696
M(Sign)           348  Pr>=|M|      0.0001
Sgn Rank       121278  Pr>=|S|      0.0001

             Quantiles(Def=5)

100% Max          80         99%        15
 75% Q3            2         95%         5
 50% Med           1         90%         3
 25% Q1            0         10%         0
  0% Min           0          5%         0
                              1%         0

Range             80
Q3-Q1              2
Mode               0
```

Exhibit 15.4

First Analysis of Benefits Survey (continued)

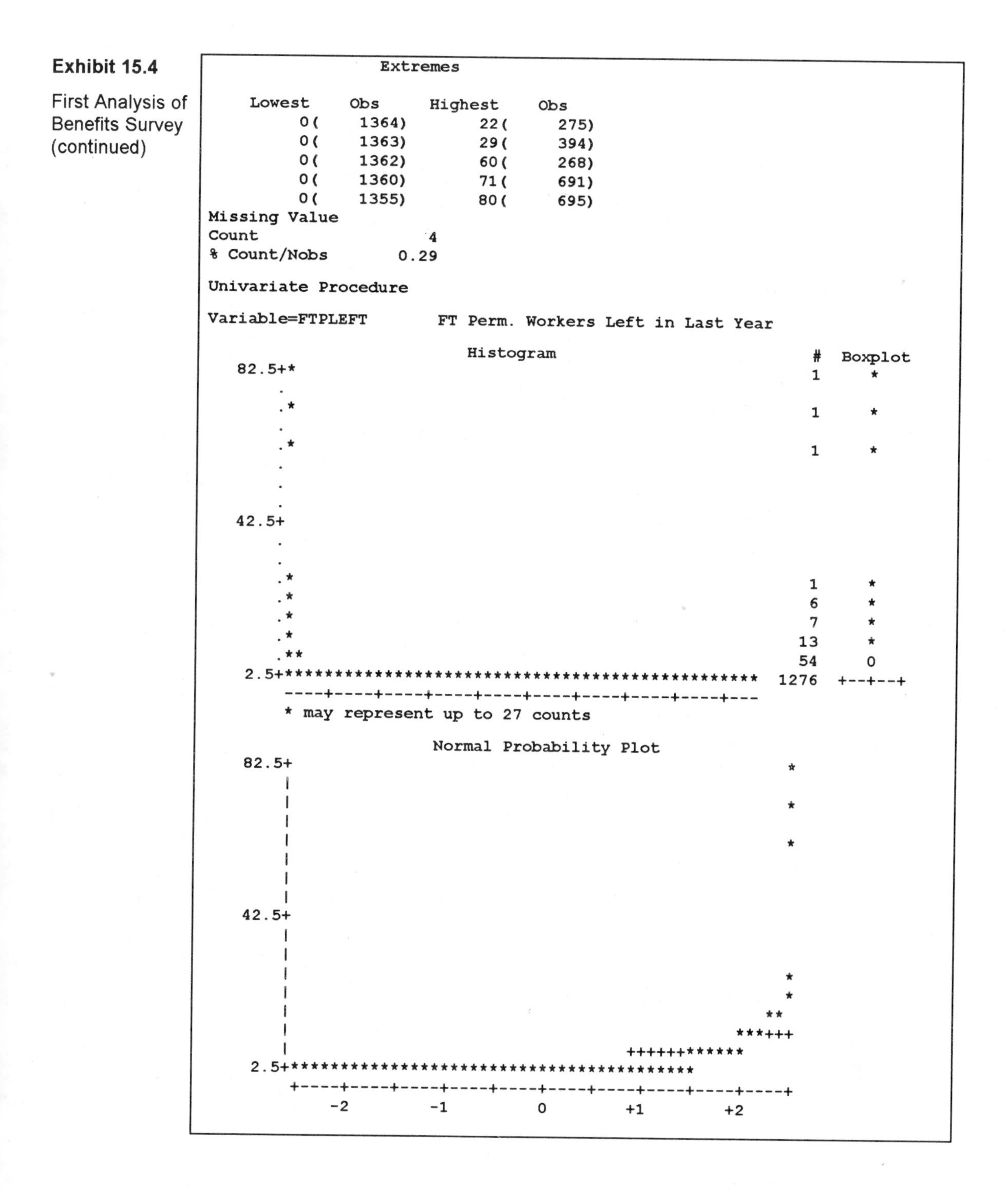

```
                    Extremes

     Lowest       Obs      Highest      Obs
          0(     1364)         22(      275)
          0(     1363)         29(      394)
          0(     1362)         60(      268)
          0(     1360)         71(      691)
          0(     1355)         80(      695)
Missing Value
Count                     4
% Count/Nobs           0.29

Univariate Procedure

Variable=FTPLEFT        FT Perm. Workers Left in Last Year

                         Histogram                          #   Boxplot
   82.5+*                                                   1      *
       .
       .*                                                   1      *
       .
       .*                                                   1      *
       .
       .
       .
   42.5+
       .
       .
       .*                                                   1      *
       .*                                                   6      *
       .*                                                   7      *
       .*                                                  13      *
       .**                                                 54      0
    2.5+************************************************ 1276   +--+--+
        ----+----+----+----+----+----+----+----+----+---
        * may represent up to 27 counts

                      Normal Probability Plot
    82.5+                                                 *
        |
        |                                                 *
        |
        |                                                 *
        |
        |
        |
    42.5+
        |
        |
        |                                                 *
        |                                                 *
        |                                               **
        |                                            ***+++
        |                                 ++++++++++++
     2.5+*****************************************
        +----+----+----+----+----+----+----+----+----+----+
            -2        -1         0        +1        +2
```

In Exhibit 15.5, we tackle the remaining tasks outlined at the beginning of this section. The initial DATA step creates a new SAS dataset. It contains an additional variable, `turnover`, a logarithmic transformation of `ftpleft`. PROC UNIVARIATE displays the distribution of this transformed variable. PROC PLOT plots it against two of the continuous variables in the analysis (`nworkers` and `hlthpct`). PROC CORR generates a correlation matrix. Finally, we use PROC REG to perform a multiple regression predicting turnover from the independent variables (`nworkers`, `nworkers`, `sicklv`, `disablty`, `lifeins`, `retire`, and `hlthpct`). The PLOT statement beneath the MODEL statement requests diagnostic plots of the residuals.

We wish to use the title given in the TITLE1 statement as the first title on all output pages. This means that we need only supply TITLE2 statements to change the second title line for the output of each procedure.

Exhibit 15.5 shows the plots produced by UNIVARIATE and REG. It is clear that our transformed dependent variable is still quite skewed. Therefore, the results of our multiple regression need to be viewed quite cautiously. We chose this example for several reasons. It illustrates using SAS for more complex regression analysis. It also emphasizes the importance of carefully examining the distributions of variables and residuals. This helps you assess the degree to which your analysis meets or violates the basic assumptions underlying regression analysis.

Exhibit 15.5

Final Analysis of Benefits Survey

Log

```
69    ***** Take Log of FTPLEFT to Deal with Skewness *****;
70    DATA temp2;
71         SET temp1 ;
72
73      *** Calculate an Employee Turnover Index        ***
74      *** Add small number to turnover to prevent ***
75      *** log of zero.  LOG10 is log base 10        *** ;
76      turnover =  LOG10(ftpleft + 1)    ;
77
78
79    ***** Examine distribution of LOG of turnover ***** ;

NOTE: Missing values were generated as a result of performing an
      operation on missing values.
      Each place is given by: (Number of times) at (Line):(Column).
      4 at 76:15   4 at 76:29
NOTE: The data set WORK.TEMP2 has 1364 observations and 13 variables.
NOTE: The DATA statement used 4.78 seconds.

80    PROC UNIVARIATE  DATA=temp2  PLOT  ;
81         VAR  turnover        ;
82
83    TITLE1 'Analysis of Influence of Employee Benefits on Turnover'
84    TITLE2 'Distribution of Log of Number of FT Employees Who Left';
85    RUN;

NOTE: The PROCEDURE UNIVARIATE used 2.2 seconds.
```

Exhibit 15.5

Final Analysis of Benefits Survey (continued)

```
86
87    ***** Check linearity for nworkers and pctpaid ***** ;
88    PROC PLOT  DATA=temp2 ;
89       PLOT  turnover * (nworkers hlthpct) ;
90
91    TITLE2 'Plot of Logged Variable Against Continuous Variables ';
92    RUN;
93
94
95    ***** Get Correlations Among All Variables *****;

NOTE: The PROCEDURE PLOT used 3.24 seconds.

96    PROC CORR  DATA=temp2 ;
97        VAR  turnover nworkers sicklv disablty lifeins retire hlthpct;
98
99    TITLE2 'Correlations Among Analysis Variables' ;
100   RUN;

NOTE: The PROCEDURE CORR used 3.45 seconds.

101
102
103   ***** Estimate regression model and check residuals ***** ;
104   PROC REG  DATA=temp2 ;
105   MODEL turnover = nworkers sicklv disablty lifeins retire hlthpct;
106
107        PLOT  r. * p.  student. * p.   student. * (nworkers hlthpct)
108             /  HPLOTS=2 VPLOTS=2 ;
109
110   TITLE2 'Multiple Regression for Log of Number of Employees Who Left';
111   RUN;

NOTE: 1364 observations read.
NOTE: 93 observations have missing values.
NOTE: 1271 observations used in computations.
```

Output Begins

```
Analysis of Influence of Employee Benefits on Turnover
Distribution of Log of Number of FT Employees Who Left

Univariate Procedure

Variable=TURNOVER

                  Moments

 N              1360  Sum Wgts       1360
 Mean       0.248666  Sum        338.1863
 Std Dev    0.298056  Variance   0.088837
 Skewness   1.305598  Kurtosis   2.259974
 USS        204.8252  CSS        120.7296
 CV         119.8616  Std Mean   0.008082
 T:Mean=0    30.7673  Pr>|T|       0.0001
 Num ^= 0        696  Num > 0         696
 M(Sign)         348  Pr>=|M|      0.0001
 Sgn Rank     121278  Pr>=|S|      0.0001
```

Exhibit 15.5

Final Analysis of Benefits Survey (continued)

```
                 Quantiles(Def=5)

 100% Max  1.908485         99%    1.20412
  75% Q3   0.477121         95%   0.778151
  50% Med   0.30103         90%    0.60206
  25% Q1          0         10%          0
   0% Min         0          5%          0
                             1%          0

 Range     1.908485
 Q3-Q1     0.477121
 Mode             0

                    Extremes

    Lowest     Obs      Highest     Obs
         0(   1364) 1.361728(     275)
         0(   1363) 1.477121(     394)
         0(   1362)  1.78533(     268)
         0(   1360) 1.857332(     691)
         0(   1355) 1.908485(     695)

Missing Value          .
Count                  4
% Count/Nobs        0.29
```

New Page of Output

```
Analysis of Influence of Employee Benefits on Turnover
Distribution of Log of Number of FT Employees Who Left

Univariate Procedure

Variable=TURNOVER

                          Histogram                       #  Boxplot
   1.95+*                                                 1     0
       .*                                                 1     0
       .*                                                 1     0
       .
       .
       .*                                                 1     0
       .*                                                 7     0
       .*                                                 6     0
       .*                                                 3     |
       .*                                                11     |
       .**                                               15     |
       .**                                               16     |
       .**                                               22     |
       .*********                                       125     |
       .                                                        |
       .************                                    164  +-----+
       .************************                        323  *-----*
       .                                                     |  +  |
       .                                                     |     |
   0.05+************************************************    664  +-----+
        ----+----+----+----+----+----+----+----+----+---
        * may represent up to 14 counts
```

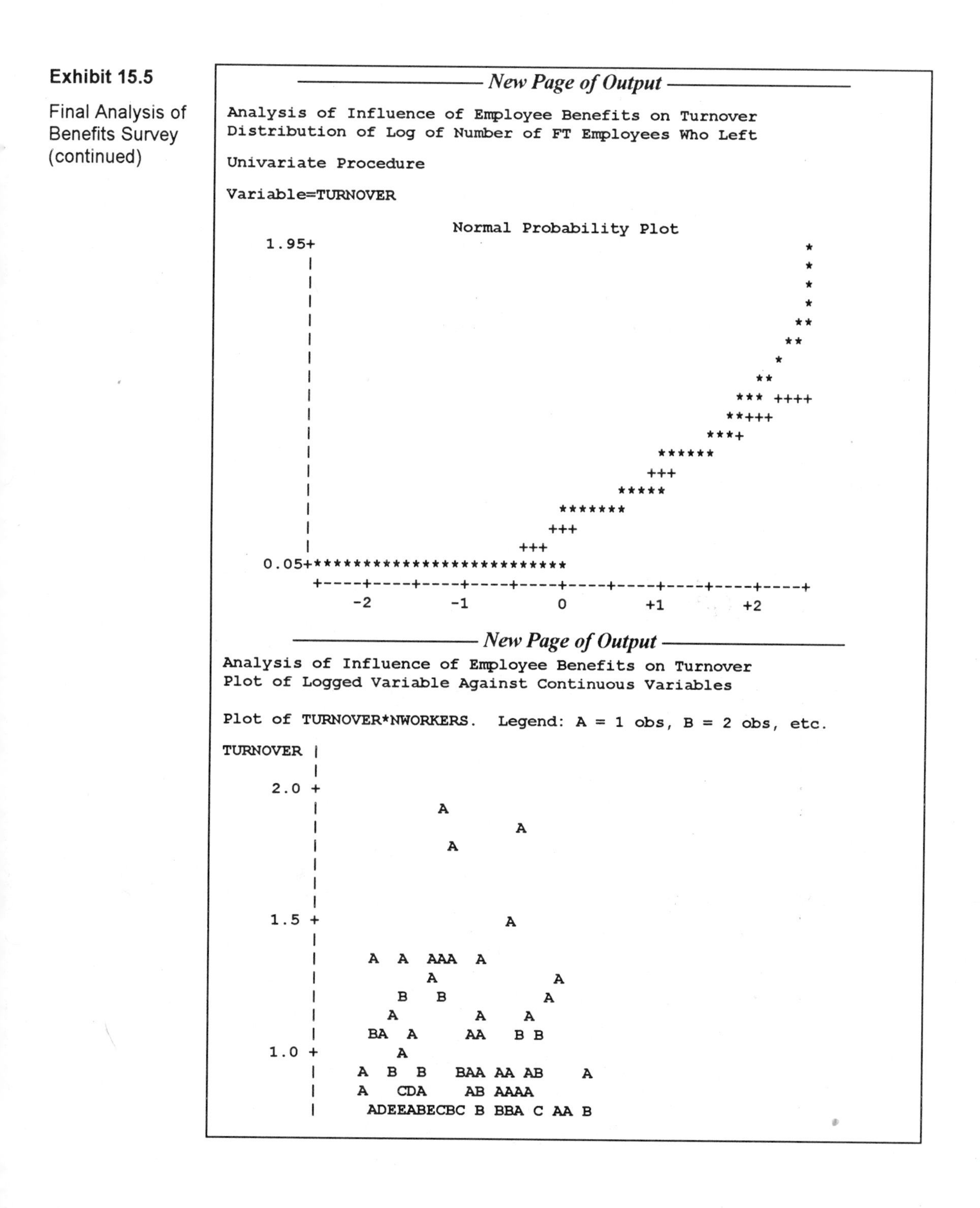

Exhibit 15.5

Final Analysis of Benefits Survey (continued)

New Page of Output

```
Analysis of Influence of Employee Benefits on Turnover
Distribution of Log of Number of FT Employees Who Left

Univariate Procedure

Variable=TURNOVER

                           Normal Probability Plot
   1.95+                                                             *
       |                                                             *
       |                                                             *
       |                                                             *
       |                                                            **
       |                                                           **
       |                                                          *
       |                                                        **
       |                                                      *** ++++
       |                                                     *****++
       |                                                   ***+
       |                                              ******
       |                                             +++
       |                                          *****
       |                                     *******
       |                                    +++
       |                                 +++
   0.05+***************************
        +----+----+----+----+----+----+----+----+----+----+
            -2        -1         0        +1        +2
```

New Page of Output

```
Analysis of Influence of Employee Benefits on Turnover
Plot of Logged Variable Against Continuous Variables

Plot of TURNOVER*NWORKERS.  Legend: A = 1 obs, B = 2 obs, etc.

TURNOVER |
         |
     2.0 +
         |            A
         |                    A
         |             A
         |
         |
         |
     1.5 +                   A
         |
         |     A  A  AAA  A
         |           A            A
         |        B   B          A
         |       A       A    A
         |     BA  A    AA   B B
     1.0 +        A
         |    A  B  B   BAA AA AB     A
         |    A   CDA    AB AAAA
         |     ADEEABECBC B BBA C AA B
```

Exhibit 15.5

Final Analysis of Benefits Survey (continued)

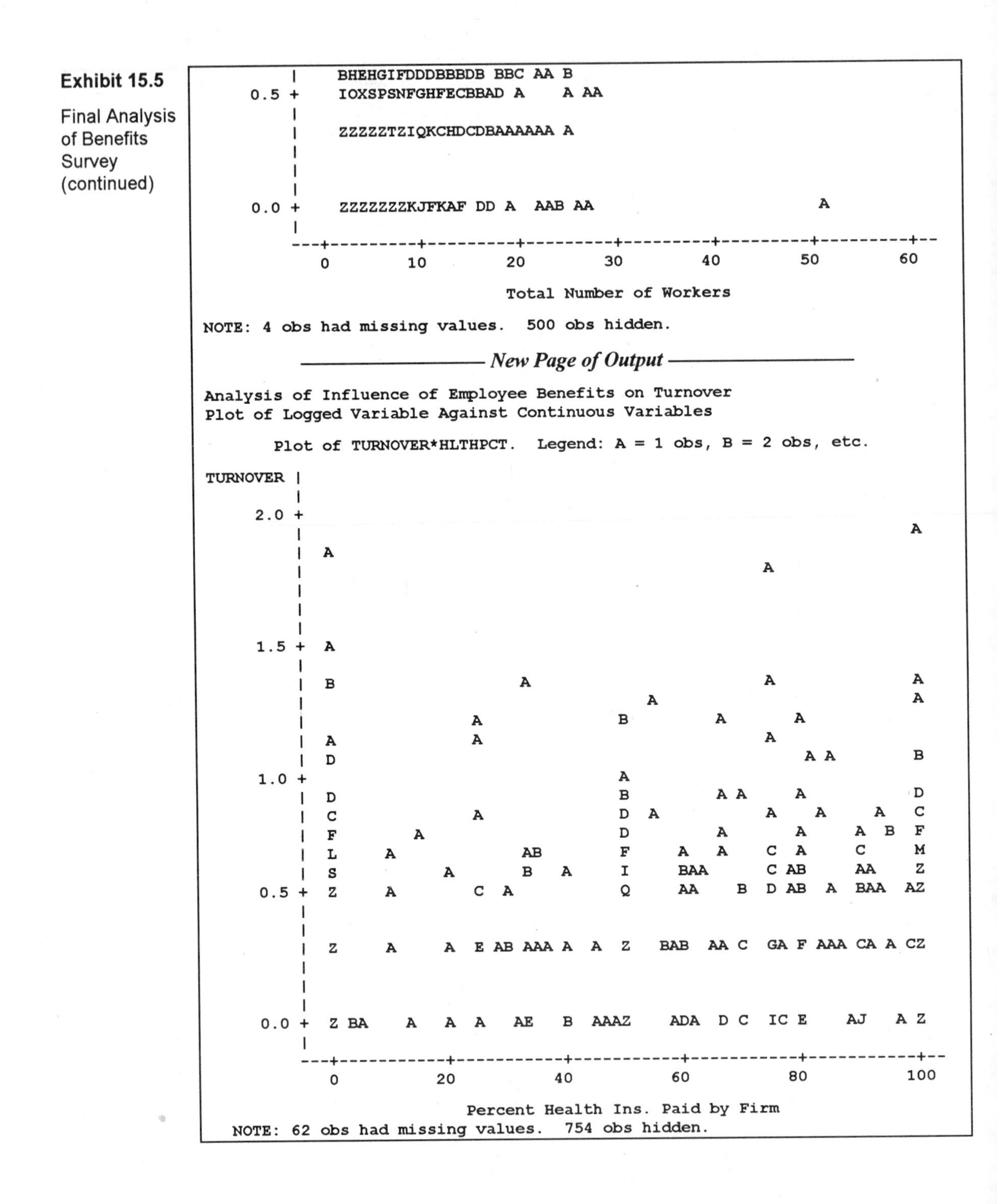

```
         |     BHEHGIFDDDBBBDB BBC AA B
   0.5 + IOXSPSNFGHFECBBAD A     A AA
         |
         |     ZZZZZTZIQKCHDCDBAAAAAA A
         |
         |
         |
   0.0 + ZZZZZZZKJFKAF DD A   AAB AA                          A
         |
      ---+---------+---------+---------+---------+---------+---------+--
         0        10        20        30        40        50        60

                                Total Number of Workers

NOTE: 4 obs had missing values.  500 obs hidden.
```

New Page of Output

```
Analysis of Influence of Employee Benefits on Turnover
Plot of Logged Variable Against Continuous Variables

       Plot of TURNOVER*HLTHPCT.  Legend: A = 1 obs, B = 2 obs, etc.

TURNOVER |
         |
   2.0 +
         |                                                              A
         | A
         |                                             A
         |
         |
         |
   1.5 + A
         |
         | B                    A                     A               A
         |                                 A                            A
         |               A             B          A      A
         | A             A                            A
         | D                                              A A          B
   1.0 +                               A
         | D                           B          A A    A             D
         | C             A             D A            A    A     A     C
         | F       A                   D          A      A     A  B  F
         | L    A             AB           F      A  A    C A      C     M
         | S         A        B   A    I      BAA      C AB      AA    Z
   0.5 + Z    A        C  A            Q      AA    B D AB  A  BAA  AZ
         |
         |
         | Z    A     A  E AB AAA A   A  Z   BAB  AA C  GA F AAA CA A CZ
         |
         |
         |
   0.0 + Z BA    A    A  A    AE   B  AAAZ   ADA  D C  IC E    AJ   A Z
         |
      ---+-----------+-----------+-----------+-----------+-----------+--
         0          20          40          60          80         100

                         Percent Health Ins. Paid by Firm
NOTE: 62 obs had missing values.  754 obs hidden.
```

Exhibit 15.5

Final Analysis of Benefits Survey (continued)

New Page of Output

```
Analysis of Influence of Employee Benefits on Turnover
Correlations Among Analysis Variables

Correlation Analysis

   7 'VAR' Variables:  TURNOVER NWORKERS SICKLV   DISABLTY LIFEINS
                       RETIRE   HLTHPCT

                            Simple Statistics

Variable               N           Mean          Std Dev            Sum

TURNOVER            1360       0.248666         0.298056     338.186292
NWORKERS            1364       6.451613         4.871030    8800.000000
SICKLV              1352       0.525148         0.499552     710.000000
DISABLTY            1341       0.210291         0.407667     282.000000
LIFEINS             1357       0.327192         0.469361     444.000000
RETIRE              1353       0.153732         0.360825     208.000000
HLTHPCT             1306      53.095712        45.124460          69343

                               Simple Statistics

Variable          Minimum        Maximum   Label

TURNOVER                0       1.908485
NWORKERS         2.000000      51.000000   Total Number of Workers
SICKLV                  0       1.000000   Offer Sick Leave
DISABLTY                0       1.000000   Offer Disability Plan
LIFEINS                 0       1.000000   Offer Life Insurance
RETIRE                  0       1.000000   Offer Retirement Plan
HLTHPCT                 0     100.000000   Percent Health Ins. Paid by
              Firm
```

New Page of Output

```
Analysis of Influence of Employee Benefits on Turnover
Correlations Among Analysis Variables

Correlation Analysis

Pearson Correlation Coefficients / Prob > |R| under Ho: Rho=0
/ Number of Observations

                                  TURNOVER  NWORKERS    SICKLV  DISABLTY

TURNOVER                           1.00000   0.42147  -0.01029   0.05778
                                    0.0      0.0001    0.7057    0.0346
                                      1360      1360      1348      1338

NWORKERS                           0.42147   1.00000   0.05883   0.03619
Total Number of Workers             0.0001   0.0       0.0305    0.1854
                                      1360      1364      1352      1341

SICKLV                            -0.01029   0.05883   1.00000   0.18639
Offer Sick Leave                    0.7057   0.0305    0.0       0.0001
                                      1348      1352      1352      1335
```

Exhibit 15.5

Final Analysis of Benefits Survey (continued)

```
DISABLTY                              0.05778   0.03619   0.18639   1.00000
Offer Disability Plan                 0.0346    0.1854    0.0001    0.0
                                        1338      1341      1335      1341

LIFEINS                               0.04249   0.13148   0.26053   0.38886
Offer Life Insurance                  0.1182    0.0001    0.0001    0.0001
                                        1353      1357      1351      1340

RETIRE                                0.00389   0.05891   0.17969   0.18601
Offer Retirement Plan                 0.8864    0.0302    0.0001    0.0001
                                        1349      1353      1347      1338

HLTHPCT                               0.00591   0.05270   0.24927   0.24845
Percent Health Ins. Paid by Firm      0.8312    0.0569    0.0001    0.0001
                                        1302      1306      1294      1284
```

New Page of Output

```
Analysis of Influence of Employee Benefits on Turnover
Correlations Among Analysis Variables

Correlation Analysis

Pearson Correlation Coefficients / Prob > |R| under Ho: Rho=0
/ Number of Observations
LIFEINS     RETIRE    HLTHPCT

TURNOVER                              0.04249   0.00389   0.00591
                                      0.1182    0.8864    0.8312
                                        1353      1349      1302

NWORKERS                              0.13148   0.05891   0.05270
Total Number of Workers               0.0001    0.0302    0.0569
                                        1357      1353      1306

SICKLV                                0.26053   0.17969   0.24927
Offer Sick Leave                      0.0001    0.0001    0.0001
                                        1351      1347      1294

DISABLTY                              0.38886   0.18601   0.24845
Offer Disability Plan                 0.0001    0.0001    0.0001
                                        1340      1338      1284

LIFEINS                               1.00000   0.25021   0.43393
Offer Life Insurance                  0.0       0.0001    0.0001
                                        1357      1352      1299

RETIRE                                0.25021   1.00000   0.20659
Offer Retirement Plan                 0.0001    0.0       0.0001
                                        1352      1353      1295

HLTHPCT                               0.43393   0.20659   1.00000
Percent Health Ins. Paid by Firm      0.0001    0.0001    0.0
                                        1299      1295      1306
```

Exhibit 15.5

Final Analysis of Benefits Survey (continued)

New Page of Output

```
Analysis of Influence of Employee Benefits on Turnover
Multiple Regression for Log of Number of Employees Who Left

Model: MODEL1
Dependent Variable: TURNOVER
Analysis of Variance

                          Sum of          Mean
Source          DF       Squares        Square       F Value       Prob>F

Model            6      19.90219       3.31703        45.217       0.0001
Error         1264      92.72420       0.07336
C Total       1270     112.62639

    Root MSE         0.27085     R-square       0.1767
    Dep Mean         0.24811     Adj R-sq       0.1728
    C.V.           109.16288

Parameter Estimates

                     Parameter       Standard    T for H0:
Variable  DF          Estimate          Error   Parameter=0    Prob > |T|

INTERCEP   1          0.096118     0.01613220         5.958        0.0001
NWORKERS   1          0.025489     0.00156350        16.303        0.0001
SICKLV     1         -0.021656     0.01607375        -1.347        0.1781
DISABLTY   1          0.045971     0.02059718         2.232        0.0258
LIFEINS    1         -0.024936     0.01950658        -1.278        0.2014
RETIRE     1         -0.011336     0.02229556        -0.508        0.6112
HLTHPCT    1      -0.000019181     0.00019122        -0.100        0.9201

                 Variable
Variable  DF       Label

INTERCEP   1   Intercept
NWORKERS   1   Total Number of Workers
SICKLV     1   Offer Sick Leave
DISABLTY   1   Offer Disability Plan
LIFEINS    1   Offer Life Insurance
RETIRE     1   Offer Retirement Plan
HLTHPCT    1   Percent Health Ins. Paid by Firm
```

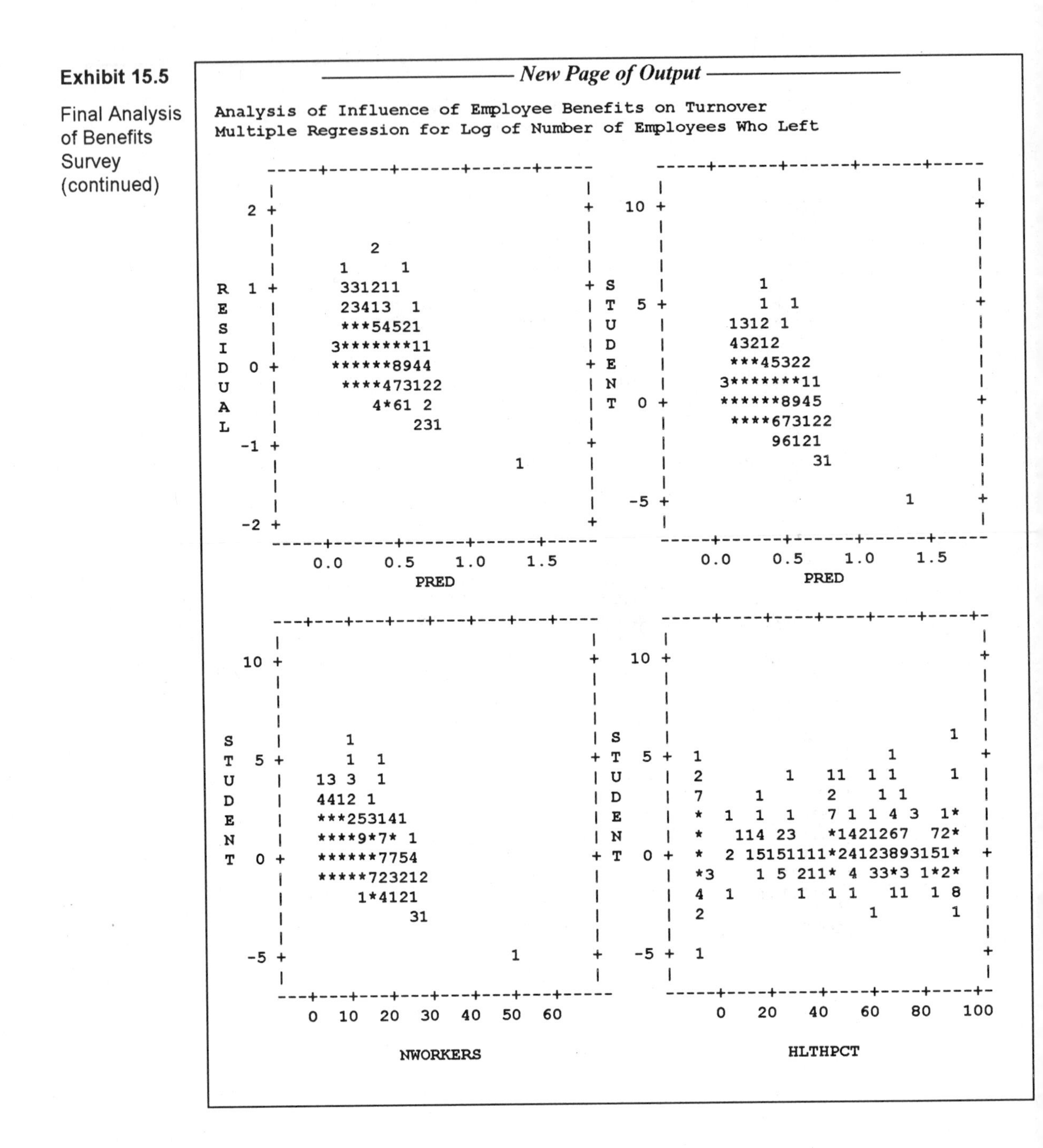

Exhibit 15.5

Final Analysis of Benefits Survey (continued)

Treating Fear of Snakes

Our last example is a rather straightforward analysis of experimental data assessing the effectiveness of several approaches to helping persons who have a fear of snakes to decrease that fear. The data are actually for a hypothetical problem posed by Maxwell and Delaney (1990) as an exercise. Imagine that a group of students who showed a fear of snakes on a test instrument were classified as either mildly or severely phobic. (Students showing no fear were not used in the study.) These students were then randomly assigned to three treatments for lessening fear of snakes: systematic desensitization, implosive therapy, or insight therapy. Eight students were assigned to each of the six possible combinations of level of phobia and treatment. At the end of their treatment, students took the Behavioral Avoidance Test to assess their fear. Higher scores meant less fear. The questions of interest to the researcher might be:

- ❑ Was there a difference in effectiveness among the treatments that depended upon the degree of fear?
- ❑ Was desensitization more effective than the other treatments?

In this example we

- ❑ Read the data from the experiment that was recorded, using abbreviations for the treatment groups and degree of fear of snakes
- ❑ Create new character variables from the codes for the abbreviations which will be useful for labeling analysis output, and perform an initial two-way ANOVA

The DATA step code shown in Exhibit 15.6 reads the data for each subject. It includes the subject's treatment condition (`trtment`), degree of fear (`fear`), and Behavioral Avoidance Test score (`batscore`). It also creates two new variables (`treat` and `phobia`), with character values that express the abbreviated codes for `trtment` and `fear` in words. We could also have defined and used output formats for this purpose, but we felt that using character-valued variables was a bit more straightforward in this instance.

PROC GLM uses the newly created variables (`treat` and `phobia`) to request a two-way ANOVA model with an interaction term that addresses the first research question. The MEANS statement obtains the means for each combination of treatment and initial level of fear.

Exhibit 15.6

First Analysis of Fear of Snakes Experiment

Log

```
1     OPTIONS  ls=75 ps=50 pageno=1 nodate nocenter;
2     **** Read data from an external file      ****;
3     **** Create character vars for analysis  ****;
4     DATA  snakes ;
5
6        INFILE  'c:\sasbook\datasets\snakes.dta' ;
7
8        INPUT  trtment $  fear $  batscore ;
9
10       IF       trtment = 'DE' THEN treat = 'Desensitization';
11       ELSE IF  trtment = 'IM' THEN treat = 'Implosion' ;
12       ELSE IF  trtment = 'IN' THEN treat = 'Insight'   ;
13
14       IF       fear = 'S'  THEN  phobia = 'Severe' ;
15       ELSE IF  fear = 'M'  THEN  phobia = 'Mild'   ;
16
17
18       LABEL  treat   ='Treatment Type'
19              phobia ='Prior Phobia Level' ;
20
21    RUN;

NOTE: The data set WORK.SNAKES has 48 observations and 5 variables.
NOTE: The DATA statement used 8.28 seconds.

22
23
24    **** Test model with main and interaction effects ****;
25    PROC GLM  DATA=snakes ;
26       CLASS  treat phobia ;
27       MODEL  batscore = treat phobia treat*phobia ;
28       MEANS  treat*phobia  ;
29    TITLE1 'Comparing Treatments for Fear of Snakes';
30    TITLE2 'Test of Main and Interaction Effects'   ;
31    RUN;

NOTE: Means from the MEANS statement are not adjusted for other terms
      in the model.  For adjusted means, use the LSMEANS statement.
```

Output Begins

```
Comparing Treatments for Fear of Snakes
Test of Main and Interaction Effects

General Linear Models Procedure
Class Level Information

Class    Levels    Values
TREAT         3    Desensitization Implosion Insight
PHOBIA        2    Mild Severe

Number of observations in data set = 48
```

Exhibit 15.6

First Analysis of Fear of Snakes Experiment (continued)

New Page of Output

```
Comparing Treatments for Fear of Snakes
Test of Main and Interaction Effects

General Linear Models Procedure

Dependent Variable: BATSCORE
                                   Sum of           Mean
Source                  DF        Squares         Square   F Value    Pr > F
Model                    5     400.854167      80.170833     10.91    0.0001
Error                   42     308.625000       7.348214
Corrected Total         47     709.479167

             R-Square            C.V.       Root MSE       BATSCORE Mean
             0.564998        23.78728        2.71076             11.3958

Source               DF      Type I SS   Mean Square  F Value     Pr > F
TREAT                 2      15.166667      7.583333     1.03     0.3652
PHOBIA                1     285.187500    285.187500    38.81     0.0001
TREAT*PHOBIA          2     100.500000     50.250000     6.84     0.0027

Source               DF    Type III SS   Mean Square  F Value     Pr > F
TREAT                 2      15.166667      7.583333     1.03     0.3652
PHOBIA                1     285.187500    285.187500    38.81     0.0001
TREAT*PHOBIA          2     100.500000     50.250000     6.84     0.0027
```

New Page of Output Begins

```
Comparing Treatments for Fear of Snakes
Test of Main and Interaction Effects

General Linear Models Procedure

Level of         Level of          -----------BATSCORE----------
TREAT            PHOBIA      N        Mean                 SD

Desensitization  Mild        8     13.2500000         1.66904592
Desensitization  Severe      8     11.1250000         3.18198052
Implosion        Mild        8     15.3750000         1.40788595
Implosion        Severe      8      6.5000000         3.89138242
Insight          Mild        8     12.8750000         1.64208056
Insight          Severe      8      9.2500000         3.37003603
```

Examination of the GLM output in Exhibit 15.6 reveals an interaction between treatment and initial level of fear significant beyond the .003 level. Thus it would seem that the effectiveness of a given treatment depended on whether a subject's level of fear was severe or moderate. A review of the means at the end of the GLM output indicates that this is the case.

The PROC GLM code shown in Exhibit 15.7 was used to further examine the pattern of interaction and address to the second research question. It examines the effect of treatment nested within level of fear and constructs contrasts to test whether desensitization differs from the other treatments for each level of fear. A glance at the last page of the GLM output shows a significant difference between desensitization and other treatments for those whose level of fear was severe. Those whose fear was moderate, however, did not have a significant difference among treatments.

Exhibit 15.7

Second Analysis of Fear of Snakes Experiment

Log

```
35   PROC GLM  DATA=snakes ;
36   CLASS  treat phobia ;
37   MODEL  batscore = treat(phobia)  ;
38   MEANS  treat(phobia)  ;
39   CONTRAST 'Des. vs. Other for M' treat(phobia)  1 -.5 -.5  0   0   0 ;
40   CONTRAST 'Des. vs. Other for S' treat(phobia)  0  0    0  1 -.5 -.5 ;
41   TITLE2 'Test of Treatment Nested Within Level of Phobia';
42   RUN;

NOTE: Means from the MEANS statement are not adjusted for other terms
      in the model.  For adjusted means, use the LSMEANS statement.
```

Output Begins

```
Comparing Treatments for Fear of Snakes
Test of Treatment Nested Within Level of Phobia

General Linear Models Procedure
Class Level Information

Class     Levels    Values
TREAT          3    Desensitization Implosion Insight
PHOBIA         2    Mild Severe

Number of observations in data set = 48
```

New Page of Output

```
Comparing Treatments for Fear of Snakes
Test of Treatment Nested Within Level of Phobia

General Linear Models Procedure

Dependent Variable: BATSCORE
                                   Sum of           Mean
Source              DF            Squares         Square   F Value     Pr > F
Model                5         400.854167      80.170833     10.91     0.0001
Error               42         308.625000       7.348214
Corrected Total     47         709.479167
```

Exhibit 15.7

Second Analysis of Fear of Snakes Experiment (continued)

```
        R-Square       C.V.        Root MSE        BATSCORE Mean
        0.564998    3.78728         2.71076              11.3958

Source                DF        Type I SS    Mean Square  F Value    Pr > F

TREAT(PHOBIA)          5      400.854167      80.170833     10.91    0.0001
```

New Page of Output

```
Comparing Treatments for Fear of Snakes
Test of Treatment Nested within Level of Phobia

General Linear Models Procedure

Level of              Level of            -----------BATSCORE----------
TREAT                 PHOBIA      N         Mean                  SD

Desensitization       Mild        8      13.2500000          1.66904592
Implosion             Mild        8      15.3750000          1.40788595
Insight               Mild        8      12.8750000          1.64208056
Desensitization       Severe      8      11.1250000          3.18198052
Implosion             Severe      8       6.5000000          3.89138242
Insight               Severe      8       9.2500000          3.37003603
```

New Page of Output

```
Comparing Treatments for Fear of Snakes
Test of Treatment Nested Within Level of Phobia

General Linear Models Procedure

Dependent Variable: BATSCORE

Contrast               DF    Contrast SS   Mean Square   F Value    Pr > F
Des. vs. Other for M    1      4.0833333     4.0833333      0.56    0.4602
Des. vs. Other for S    1     56.3333333    56.3333333      7.67    0.0083
```

Recap

Closing Comment

These examples just begin to scratch the surface of the possible uses of SAS for statistical data management and analysis. We have used only a small number of the tools reviewed in this book, and they are only a subset of what's available in the SAS System as a whole. Our hope is that we have helped you to begin fairly quickly to use SAS for statistical analysis and basic data management. We also hope you are now equipped to proceed to apply what you've learned to your own problems and will be able to pursue more advanced applications of SAS using the references we have cited throughout this book.

References

Davis, James and Tom W. Smith, *General Social Surveys, 1972–1993, Cumulative Codebook.* Chicago: National Opinion Research Center, producer,1993. Storrs, Connecticut: Roper Public Opinion Research Center, distributor, 1994.

McLaughlin, Catherine G., *Small Business Benefit Study (SBBS)*, 1990. Ann Arbor: University of Michigan Survey Research Center, producer, 1990. Ann Arbor: Interuniversity Consortium for Political and Social Research, distributor, 1994.

Maxwell, Scott E. and Harold D. Delaney, *Designing Experiments and Analyzing Data: A Model Comparison Perspective*. Belmont, CA: Wadsworth Publishing Co., 1990.

Using the Display Manager

Most of this book's chapters emphasize the SAS System's tools and how to put them together in a program. Just how they are actually put in motion and run is not discussed in great detail. In this appendix, we illustrate the use of the most common interactive interface to the SAS System. The Display Manager is a windowing environment that allows you to develop and run SAS programs. In this appendix, we discuss the layout of the Display Manager's windows, illustrate its power by developing an example, and present some of its more useful commands.

Bear in mind that the appearance of the screens shown in the exhibits may not match that of the screens on your system. The exhibits show the Display Manager running in a DOS/Windows operating system running SAS Version 6.10. Other systems and versions will differ in appearance. However, the names of the windows and the basic Display Manager commands we illustrate are the same across all versions of SAS starting with Version 6.08.

Display Manager Background

Part of the appeal of the Display Manager to many SAS users is its ability to let you enter program statements and data, run them, and be able to view the results immediately. This immediacy is enhanced by the Display Manager's ability to let you store the statements you ran and edit them later in the same session and, possibly, in a later one. The Display Manager can become an important tool to help you develop programs quickly.

Nearly every SAS installation has its own, slightly different, way to invoke the Display Manager. On many systems, you simply type `sas` at the system prompt (the TSO "READY", DCL "$", or DOS "C:\> ", for example). On most UNIX systems (Sun, IBM RISC 6000, Hewlett-Packard, and so on), you *must* type `sas` in lowercase. In graphical environments, such as Windows or OS/2, there may be a SAS icon. Double-clicking on the icon will begin execution of SAS using the Display Manager. Once SAS completes its setup, it displays a screen similar to that shown in Exhibit A.1.

Exhibit A.1

Standard Display Manager Screen Displayed at SAS System Startup

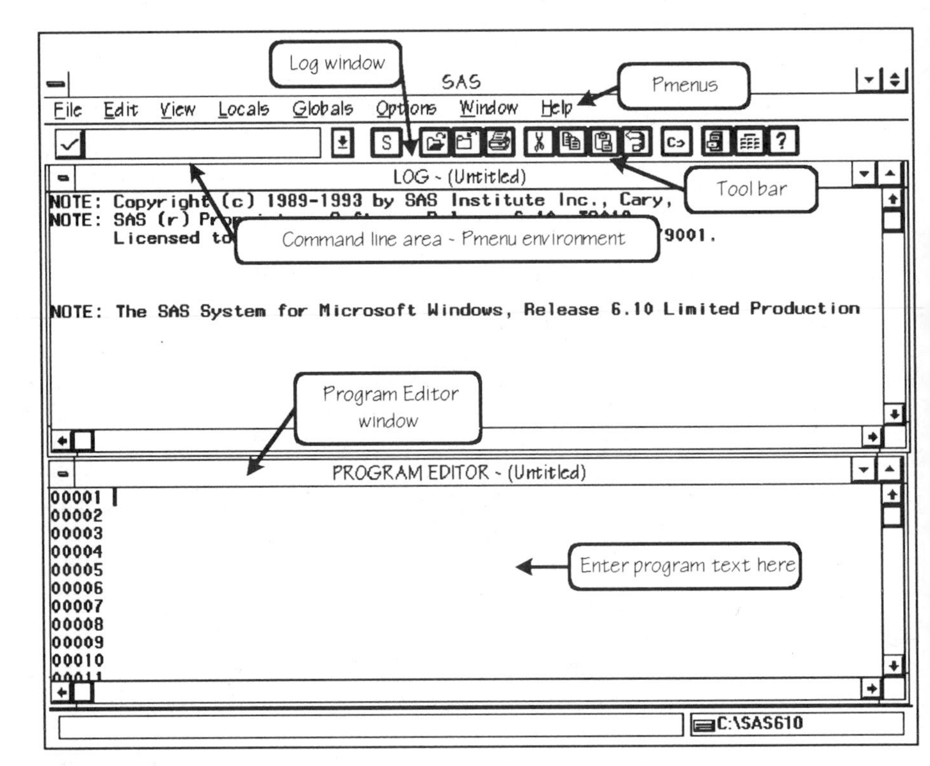

The exhibit contains most of the elements of the Display Manager environment. The screen is divided into several windows, or panels. In this case, we see a Log window and a Program Editor window. Each has a specific function in the interactive SAS environment. The Program window allows entry of SAS program statements and data lines. The Log window contains the Log for any submitted program statements, just like the Log *file* in noninteractive environments. Other windows include Options (to set SAS options, discussed in Chapter 11), Output (containing output from the SAS procedures), Keys (used for function-key assignments), and many others. For the sake of comparison, Exhibit A.2 shows the initial screen in an OpenVMS environment, similar to those of VAX operating systems.

Exhibit A.2

Startup Screen in an OpenVMS Environment

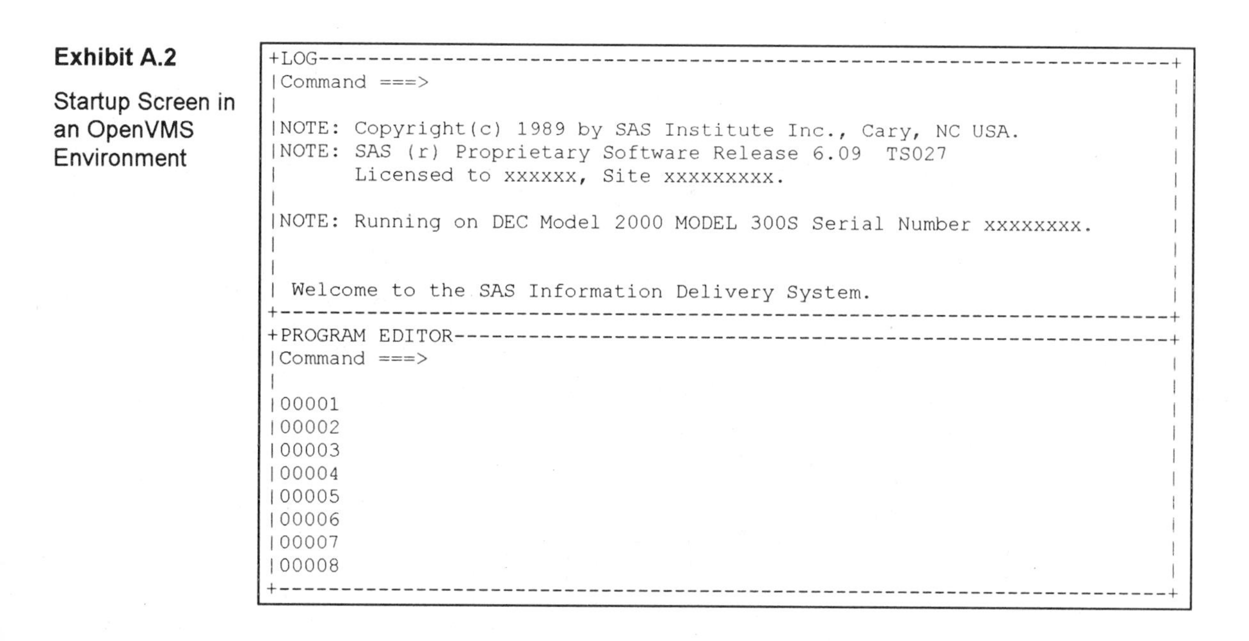

```
+LOG-----------------------------------------------------------------------+
|Command ===>                                                               |
|                                                                           |
|NOTE: Copyright(c) 1989 by SAS Institute Inc., Cary, NC USA.               |
|NOTE: SAS (r) Proprietary Software Release 6.09  TS027                     |
|      Licensed to xxxxxx, Site xxxxxxxxx.                                  |
|                                                                           |
|NOTE: Running on DEC Model 2000 MODEL 300S Serial Number xxxxxxxx.         |
|                                                                           |
|                                                                           |
| Welcome to the SAS Information Delivery System.                           |
+---------------------------------------------------------------------------+
+PROGRAM EDITOR-------------------------------------------------------------+
|Command ===>                                                               |
|                                                                           |
|00001                                                                      |
|00002                                                                      |
|00003                                                                      |
|00004                                                                      |
|00005                                                                      |
|00006                                                                      |
|00007                                                                      |
|00008                                                                      |
+---------------------------------------------------------------------------+
```

Now that you have started the Display Manager, what next? You can enter and run a program. But first, let's identify some of the ways to issue Display Manager commands. There are two types of command environments in the Display Manager; both are usually available in most versions of SAS. The first environment is often used in graphical (Windows, OS/2, Motif, X-Windows) systems. It uses pull-down menus, often called *Pmenus*, which group commands into a menu tree structure. Double-click on a menu item to perform an action or to be prompted for more information. Exhibit A.3 illustrates Pmenus for the Display Manager in SAS Version 6.10.

In the exhibit we double-clicked on File, then on Open. The system prompts us for the name of a SAS program file to be brought into the Program Editor. When we enter the file name and click on OK, the file is read into the Editor window.

The content of a Pmenu varies according to the window. Items that are appropriate in the Program Editor (line numbering and other editing commands, for example) are not appropriate in others (the Log or Options windows). The Display Manager is always aware of which window is active and presents only options that are appropriate for that window. Pmenus are an excellent way for a beginner to snoop around the SAS System and see what's available. They are also a good way for the occasional SAS user to review Display Manager commands.

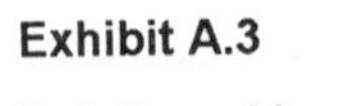

Exhibit A.3

Pull-Down Menus

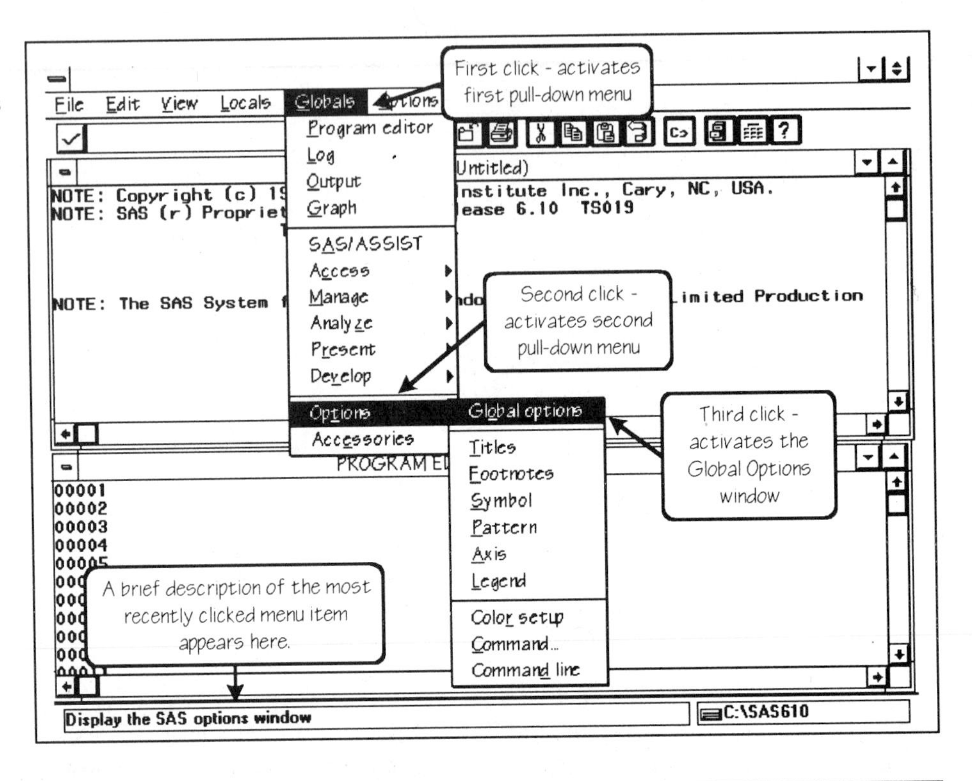

Exhibit A.4

Command Line Environment

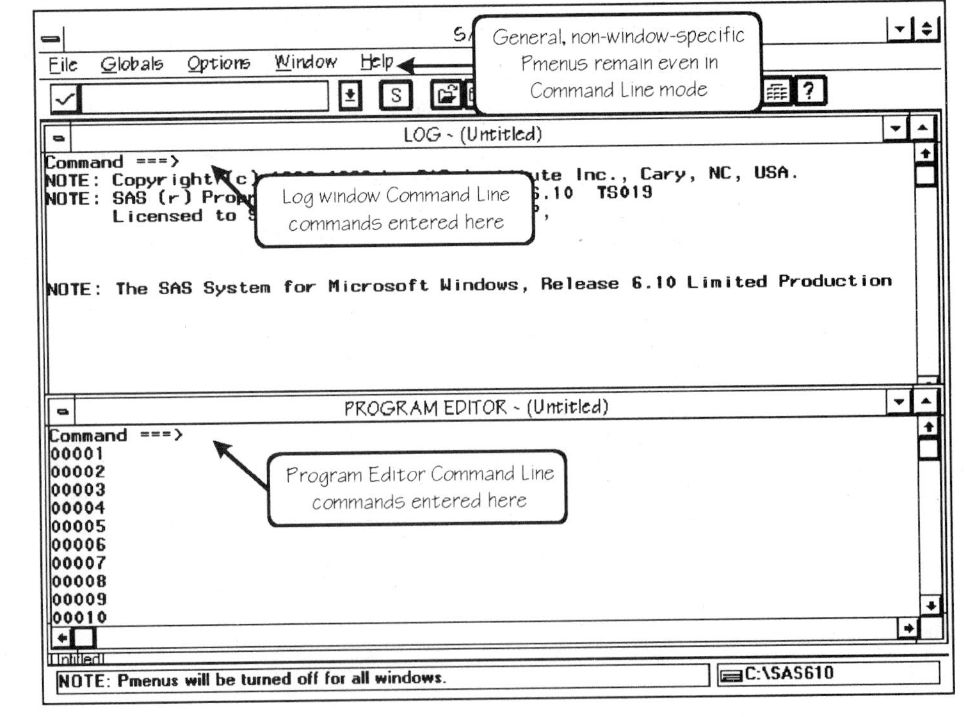

The other way to use the Display Manager is the Command Line environment. Here, rather than use menus to issue commands, you issue commands directly by typing them on a command line, located at the top of most windows. You gain speed because you can issue more than one command at a time, something you cannot do with Pmenus. Although working in Command Line mode is usually faster than Pmenus, it carries a price: you need to remember many commands. A Command Line screen is presented in Exhibit A.4.

Syntax

It's difficult to display on a page a process that is best viewed in person or at least on video. A few syntax conventions are used in the discussion that follows. These facilitate the potentially awkward page-to-screen transition. We present both Command Line and Pmenu versions of commands whenever possible. The Command Line syntax is straightforward:

```
CMD option ...
```

where *CMD* is the name of the command and *option* represents one or more items used by the command. A semicolon (;) may be used to separate multiple commands entered on the Command Line. Here are some examples of Command Line commands:

```
include 'testprog.sas'
file 'newprog.sas'
log;clear;program
```

Our syntax for Pmenu items identifies the menu item names, separating them with a vertical bar. Consider the following:

```
locals | recall
```

This means move to the Pmenu, select `locals`, and then select `recall`.

A note about selecting items from Pmenus: There are several ways to select an item, and they vary according to the operating system. Generally, you can position the cursor at the Pmenu bar and use the Tab key to move from item to item. Select an item by pressing Return or Enter. Move the cursor to the pull-down item you want and press Return or Enter again. The other method for selection usually works only in graphical environments: a double-click of the mouse on the desired Pmenu item will activate it.

A Sample Display Manager Session

By now, you know the advantages of using the Display Manager, can distinguish between its operating modes, and are able to follow the syntax descriptions. We will demonstrate the Display Manager by working through a sample session. Some commands are discussed here. They are also noted in the syntax summary at the end of this chapter. We use Command Line mode, but also present the equivalent Pmenu commands when appropriate.

Entering the Program

Let's start by entering a short program in the Program Editor. This is presented in Exhibit A.5.

Exhibit A.5

Short Program Entered in Program Editor Window

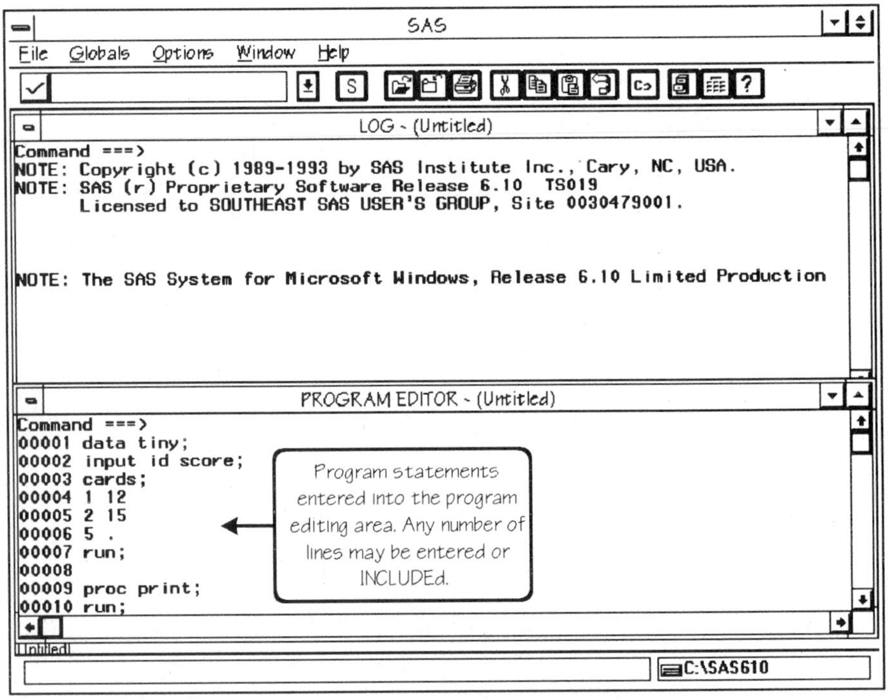

SAS

File Globals Options Window Help

LOG - (Untitled)

```
Command ===>
NOTE: Copyright (c) 1989-1993 by SAS Institute Inc., Cary, NC, USA.
NOTE: SAS (r) Proprietary Software Release 6.10  TS019
      Licensed to SOUTHEAST SAS USER'S GROUP, Site 0030479001.


NOTE: The SAS System for Microsoft Windows, Release 6.10 Limited Production
```

PROGRAM EDITOR - (Untitled)

```
Command ===>
00001 data tiny;
00002 input id score;
00003 cards;
00004 1 12
00005 2 15
00006 5 .
00007 run;
00008
00009 proc print;
00010 run;
```

Program statements entered into the program editing area. Any number of lines may be entered or INCLUDEd.

Untitled

C:\SAS610

Entering text in the editing area is simple, especially for users of editors like ISPF or XEDIT. Just type, then use the Return or Enter key to go to a new line. The Backspace and Insert keys are usually recognized by the Display Manager, so you can treat the editing area as you would any other text processing environment.

Running the Program

Later on, we'll discuss some of the commands used to edit the program. For now, let's just run the program and see what happens. Note the RUN statements in the program. These are very important in the Display Manager! RUN tells the SAS System that the statements submitted for execution are actually ready to be processed. Submitting a program without RUN is rather like telling the Display Manager, "Here's my program. Don't do anything with it. Just hold on to it until I'm ready." Submitting a program with a RUN statement tells SAS, "OK, let's see how it runs."

To send the program out of the Program Editor and execute it, use the SUBMIT command:

Command Line:	`SUBMIT`
Pmenus:	`locals | submit`

If the program runs successfully and produces output from the PRINT procedure, the Display Manager will change the active window to Output, as shown in Exhibit A.6.

Exhibit A.6

Output Window Showing Results of PRINT Procedure

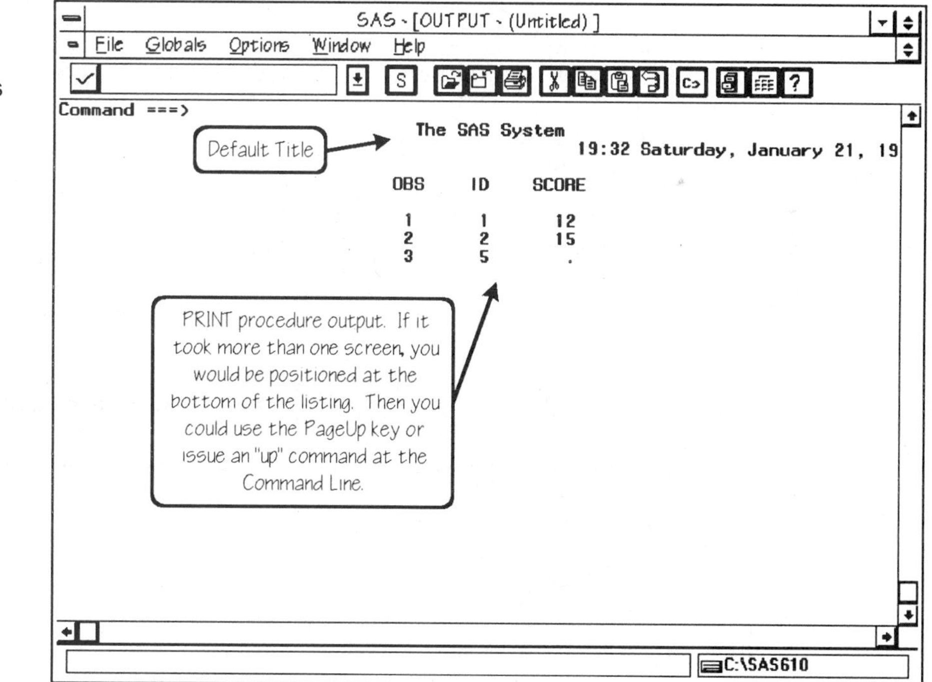

Moving Through the Output

Here in Output, as in all Display Manager windows, you have a number of options for moving through the window. In graphical environments you can use the slider bars and arrows (see Exhibit A.1) to move up, down, left, and right. If you use a PC-compatible keyboard, the PageUp and PageDown keys will move you vertically through the window. VAX environments support use of the Next and Prev keys. You can issue several movement commands when you are using the Command Line mode of the Display Manager:

```
DOWN amount
UP amount
RIGHT amount
LEFT amount
```

You do not have to specify `amount`. Simply entering the direction is sufficient. `amount` may be either `max` (`down max` moves as far down as possible) or a number of lines (`up 5` moves the display up five lines). Two other commands are available:

```
TOP
BOTTOM
```

These are the equivalents of `up max` and `down max`, respectively.

Revising the Program

Even experienced programmers and statisticians don't get their programs correct or in final form on the first try. One of the Display Manager's most helpful features is its ability to let you revise your program and resubmit it. To revise a previously run program, first move to the Program Editor window:

```
Command Line:    PROGRAM
Pmenus:          window | program editor
```

The Program Editor window will be blank. Now retrieve the last statements you SUBMITted:

```
Command Line:    RECALL
Pmenus:          locals | recall text
```

The program in Exhibit A.5 is back in the Program Editor window. Suppose we want to add a VAR statement and a title to the PRINT procedure. We need to insert two lines into the program. This is easily done by entering an insert command (the letter I) in the numbered area to the left of the line containing the

PRINT procedure. The "I" command is the first example we've seen of a *line command.* (Till now, all others have been Command Line or Pmenu commands.) Exhibit A.7 shows how we used the command to insert two blank lines into the editor area. We will see many other examples of line commands in the Syntax Summary at the end of this appendix.

Exhibit A.7

Using a Line Command to Insert Blank Lines in the Program Editor Window

```
SAS
File  Globals  Options  Window  Help

LOG - (Untitled)
Command ===>

NOTE: The data set WORK.TINY has 3 observations and 2 variables.
NOTE: The DATA statement used 2.18 seconds.

7     run;
8
9     proc print;
10    run;

NOTE: The PROCEDURE PRINT used 0.77 seconds.

PROGRAM EDITOR - (Untitled)
Command ===>
00001 data tiny;
00002 input id score;
00003 cards;
00004 1 12
00005 2 15
00006 5 .
00007 run;
00008
i2    proc print;
00010 run;

C:\SAS610
```

"i2" entered in the line number area requests insertion of two lines between the PROC and RUN lines.

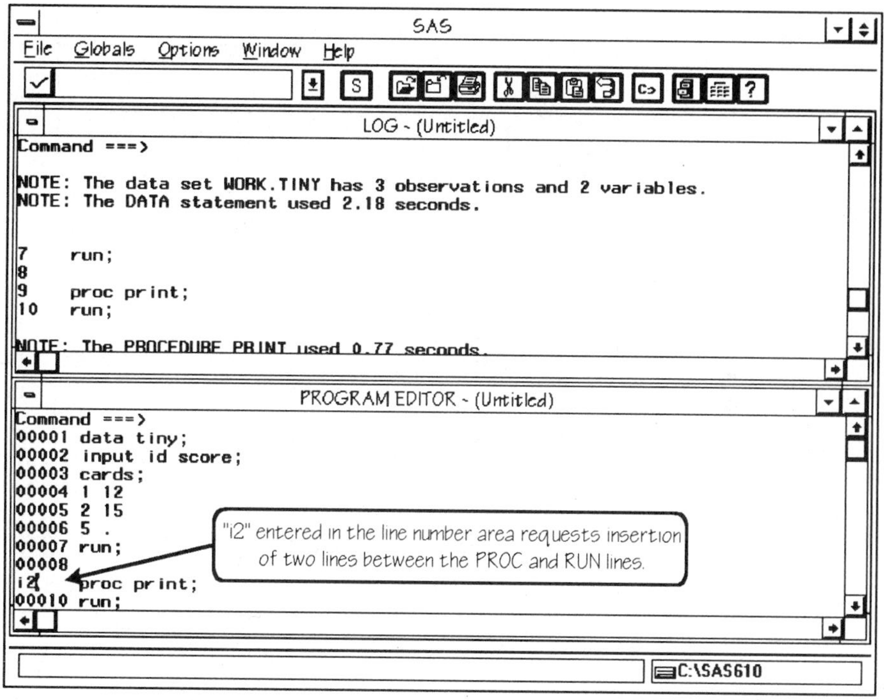

We can enter new statements, use SUBMIT to rerun the program, and use the OUTPUT and LOG commands to move to the windows containing the revised program's output and Log.

Saving Your Work

This program development cycle — enter code, submit, review, recall, edit, resubmit — is fine for the duration of the program, but what if you want to save your work for use in a later SAS session? That is, how do you save your work if you want to exit the Display Manager? Move to the Program Editor window, then use RECALL to bring the program into the editing area. Then issue a FILE command:

```
Command Line: FILE 'name'
Pmenus:       file | save as | write to file | name
```

In the above, `name` identifies the file that will contain the program statements currently in the editing area. If `name` already exists, the Display Manager issues a warning and asks for permission to overwrite the existing version. Examples of the FILE command are presented in Exhibit A.8.

Exhibit A.8

Sample FILE Commands

System	FILE Command	Comments
DOS, Windows	`file 'testprog.sas'`	Writes file TESTPROG.SAS the current directory.
	`file 'c:\code\testprog.sas'`	Writes file TESTPROG.SAS the C drive directory \CODE.
VMS	`file '[jdoe.code]testprog.sas'`	Writes file TESTPROG.SAS the directory JDOE.CODE.
	`file '[]testprog.sas'` `file 'testprog.sas'`	Each of these commands writ file TESTPROG.SAS in the current directory.
UNIX	`file 'testprog.sas'`	Writes file testprog.sas (note lowercase!) in the current directory.
	`file '~/project/testprog.sas'`	Writes file testprog.sas in a s directory of your home direct
MVS	`file 'testprog.sas'`	Writes file TESTPROG.SAS. complete file name will be *userid*.TESTPROG.SAS.
	`file 'saspgms.cntl(testprog)'`	Writes member TESPROG ir partitioned dataset (PDS) SA CODE.CNTL.

Leaving SAS

The FILE command may also be issued from most Display Manager windows, including Log and Output, so you can use it to save the Log and output for your program. Once you have saved what you need from the SAS session, you can exit by entering

```
Command Line:    BYE
Pmenus:          file | exit SAS
```

In some systems, SAS will ask you to verify that you want to exit. Once you respond `yes`, SAS and the Display Manager terminate and you are returned to your operating system.

Retrieving Your Work

The Display Manager lets you retrieve your work as well as save it. The INCLUDE command copies the contents of a file into the editing area. Its syntax is nearly identical to that of FILE, descibed in the preceding section:

Command Line:	`INCLUDE 'name'`
Pmenus:	`open | read file | name`

name is specified exactly the same way as in the FILE command. Once you successfully include a file, you can submit it, modify it, and file it. INCLUDE and FILE are, in a sense, the Display Manager's links to text files. Windows, OS/2, and some UNIX system users may also use dialog boxes to navigate the directory structure.

Customizing Your Environment with the KEYS Window

The Display Manger provides you with a way to assign Command Line commands to function keys or to various key combinations. For example, instead of continually typing SUBMIT on the Command Line, you can assign the SUBMIT command to, say, the F1 key. Once this is done, you can simply press F1 to issue the command.

To see which commands are currently assigned to the function keys, issue the KEYS command:

Command Line:	`KEYS`
Pmenus:	`help | keys`

This brings up the KEYS window, shown (Windows format) in Exhibit A.9. It is a typical Display Manager window: use the PageUp and PageDown keys or the DOWN and UP commands to scroll through the window.

Exhibit A.9

KEYS Window

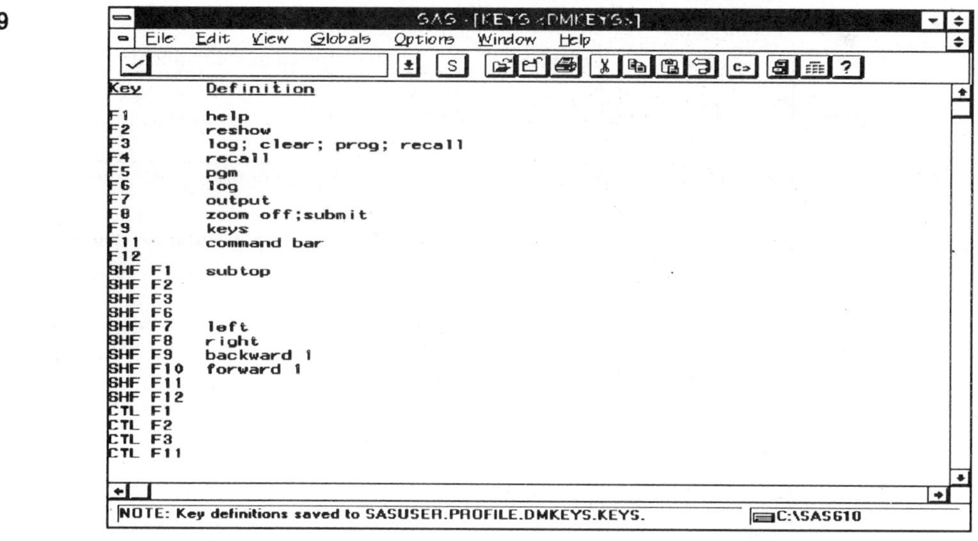

Notice the definition of the F3 key: It moves to the Log window (`log`), erases its contents (`clear`), moves to the Program window (`prog`), and fills it with the most recently submitted program statements (`recall`). You can define similar combinations as you work with the Display Manager. These are great time-savers. Remember, however, to save the key assignments for use in later sessions:

```
Command Line:    WSAVE
Pmenus:          file | save
```

Syntax Summary

This section summarizes some of the more commonly used commands. Keep in mind that there are *many* others. The best way to learn about the Display Manager is to use it, and to read the help files that come with the SAS System (see the description of the Help command). Another useful reference is the SAS Institute publication *SAS Language: Reference, Version 6, First Edition.* Chapters 17 through 19 discuss the Display Manager in detail.

Command Line Commands

Exhibit A.10 presents Command Line commands (and their Pmenu equivalents where appropriate). If you are using a graphical environment such as Windows or OS/2, you should remember that your system's windowing commands (minimize, maximize, close, and so on) are also available.

Exhibit A.10

Command Line Commands

Pmenu Equivalent	Command	Description and Comments	Command Line Example
	`?`	Retrieve commands issued from Command Line. Issue *n* times to retrieve the *n*th previous command.	`?`
	`bottom`	Move to the bottom of the current window's text.	`bottom`
`edit | clear`	`clear`	Clear the contents of the current window.	`clear`
	`down`	Scroll down in the current window.	`down max` `down 5`
`file | save as`	`file`	Save the contents of the current window in a file.	`file 'testprog.sa`
`help | extended help`	`help`	Display initial help menu or get help for a specific topic.	`help` `help print`
`file | open`	`inc` `include`	Include a file in the editing area of the Program Editor.	`inc 'testprog.sa`

Exhibit A.10

Command Line Commands (continued)

Pmenu Equivalent	Command	Description and Comments	Command Line Usage
help | keys	keys	Display settings of function keys and various Ctrl-key and Alt-key sequences.	keys
	le left	Scroll the window left. Optionally, specify the number of columns to move.	left max left left 20
edit | options | numbers	num	Turn line numbering in the Program Editor screen on or off.	num num on num off
globals | options | global options	options	Move to the Options window. You can review and, optionally, change SAS System settings.	options
globals | options | commands | pmenu	pmenu	Turn Pmenus on and off. If entered without on or off parameters, it switches states. That is, if Pmenus were on and pmenu were entered, the menus would be turned off.	pmenu pmenu on pmenu off
locals | recall	recall	Copy the most recently SUBMITted statements into the Program Editor's editing area.	recall
edit | options | reset	reset	Clear all line commands from the Program Editor window. Useful when you get confused or want to back out of a move, delete, or copy.	reset
	ri right	Scroll the window right. Optionally, specify the number of columns to move.	right max right right 20
locals | submit	sub submit	Submit the contents of the Program Editor's editing area for execution.	submit
	top	Move to the top of the current window.	top
edit | undo	undo	Remove editing changes from the Program Editor's editing area. Enter as many times as needed to undo changes.	undo
	up	Scroll up in the current window.	up max up 5
view | save attributes	wsave	Save the current window's settings. Attributes such as window size, location, and line numbering are saved.	wsave
globals | options | commands	x	Send a command to the operating system or return temporarily to the operating system. The Display Manager will tell you how to return to SAS.	x 'dir *.sas' x
	z zoom	Enlarges the current window to fill the screen or SAS window. This option expedites review of lengthy windows. If entered without on or off parameters, it switches to zoomed or unzoomed.	z z on z off

Line Commands

Exhibit A.11 describes the more commonly used line commands in the Program Editor window. To use these commands, the NUM command (described in Table A.2) should be used to turn numbering on. See the discussion following this Table for an illustration of how to use the line commands.

Exhibit A.11

Line Command Summary

Command	Description and Comments	Usage
a	Identifies the destination line for a copy or move command. The copied or moved line(s) should be placed *after* the line containing the a command.	a0001
b	Similar to a, but places lines *before* the line containing the b command.	b0005
c	Copy one, several, or a range of lines to a location specified by the a or b line command. The copied lines are left in their original locations. cc requests a "block copy," requiring another cc line command to identify the range of lines to be copied.	c0001 c3 01 cc 01 cc 09
d	Delete one, several, or a range of lines. dd requests a "block delete," requiring another dd line command to identify the range of lines to be deleted.	d0001 d5 03 dd 01 dd 08
m	Move one, several, or a range of lines to a location specified by the a or b line command. The lines are deleted from their original location. "mm" requests a "block move," requiring another mm line command to identify the range of lines to be moved.	m0002 m5 02 mm 10 mm 20
r	Repeat one, several, or a range of lines. Use a number next to the r or rr command to repeat the lines *n* times.	r0003 r10 9 rr 08 rr 20 rr3 7 rr 20

Exhibits A.12 and A.13 demonstrate the use of some of these commands. They are presented in a "before-and-after" format. Exhibit A.12 illustrates use of a block delete command. Exhibit A.13 uses the block copy (CC) command and the A command to duplicate a range of lines.

Exhibit A.12

Impact of Block Delete Line Commands

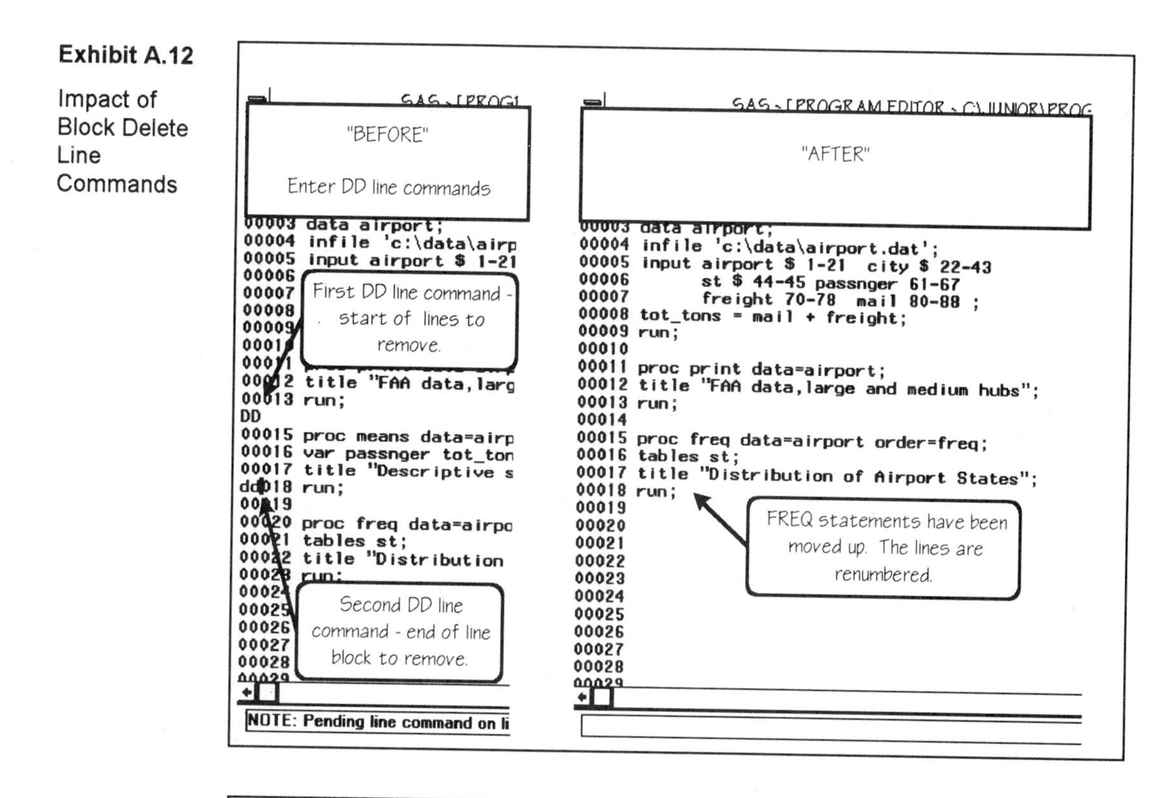

Exhibit A.13

Impact of Block Copy Line Commands

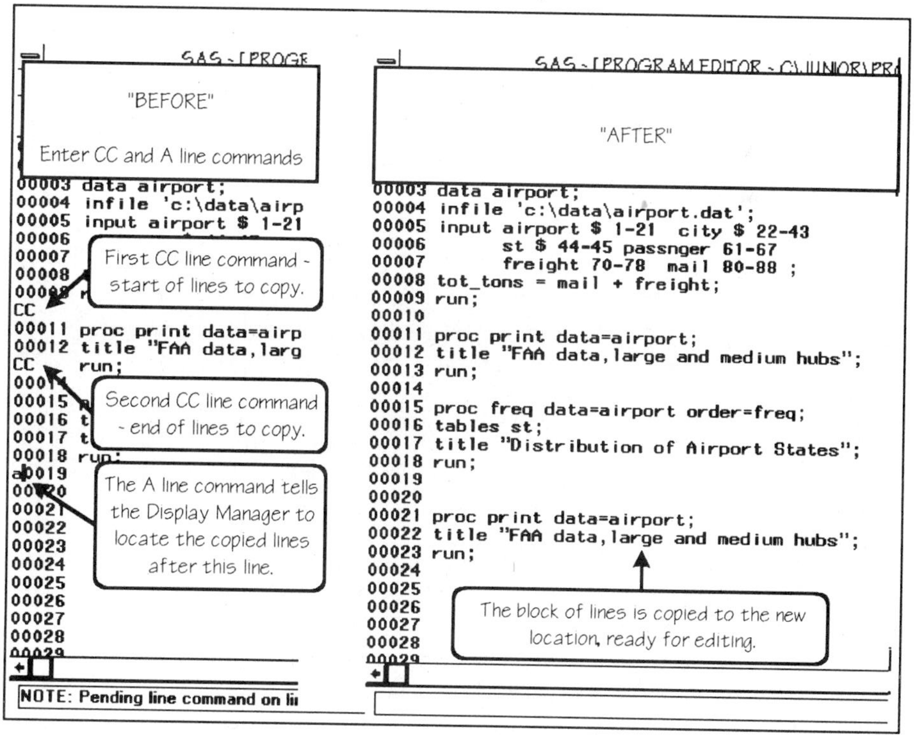

"BEFORE"

Enter CC and A line commands

```
00003 data airport;
00004 infile 'c:\data\airp
00005 input airport $ 1-21
00006
00007
00008
00009 r
CC
00011 proc print data=airp
00012 title "FAA data,larg
CC    run;
00014
00015 p
00016 t
00017 t
00018 run;
a0019
00020
00021
00022
00023
00024
00025
00026
00027
00028
NOTE: Pending line command on li
```

First CC line command - start of lines to copy.

Second CC line command - end of lines to copy.

The A line command tells the Display Manager to locate the copied lines after this line.

"AFTER"

```
00003 data airport;
00004 infile 'c:\data\airport.dat';
00005 input airport $ 1-21   city $ 22-43
00006        st $ 44-45 passnger 61-67
00007        freight 70-78  mail 80-88 ;
00008 tot_tons = mail + freight;
00009 run;
00010
00011 proc print data=airport;
00012 title "FAA data,large and medium hubs";
00013 run;
00014
00015 proc freq data=airport order=freq;
00016 tables st;
00017 title "Distribution of Airport States";
00018 run;
00019
00020
00021 proc print data=airport;
00022 title "FAA data,large and medium hubs";
00023 run;
00024
00025
00026
00027
00028
```

The block of lines is copied to the new location, ready for editing.

Resources

B

No single SAS book can claim to be exhaustive, site-specific, and physically portable. The SAS System is simply too large, and its range of applications too great, to be discussed in appropriate detail in one volume. Add to this site-specific computing issues such as how to print Logs and listings, whether the Display Manager is available, and so on, and you can readily see why you will soon need other resources to learn more about how to use SAS.

This appendix addresses that need. It identifies a variety of "liveware," print, and electronic resources, many of them available at little or no cost. It's unlikely that a single source will do the the trick for you — choose the resources that work best for you, and *don't be shy about seeking help.* Everyone, after all, was a beginner at some point in his or her career!

Person-to-Person

People who prefer to get help from "liveware" have a number of options.

Help Desks

Also referred to as *user services* or *information centers*, these groups provide walk-in and telephone assistance to computer users in an organization. They are especially valuable when you are trying to deal with site-specific tasks such as the command to run SAS interactively, special procedures for running SAS in background (or "batch") mode, printer names and locations, and the hours for computer labs. Now is the time to write the location and phone number of yours on the inside front cover of this book!

"Power Users"

These are the students or workers in an organization who appear to know everything and are willing to give advice. Word of mouth is usually the best way to identify these people. They are usually more willing to offer their help if the problem is complex and challenging. Don't be afraid to ask them questions, but don't abuse their good will.

User Groups

SAS user groups are found worldwide and come in all sizes. They have periodic meetings where people present papers, discuss programming problems and techniques, and "network." These groups often maintain electronic bulletin boards, publish newsletters, and make job referrals. Membership is usually free or available at a nominal charge.

Many companies have in-house user groups. Dozens of groups exist at the city and state levels. Larger, regional groups have been formed throughout the United States. Their annual meetings draw up to 1,000 attendees. At the highest level of geography is the SAS Users Group International (SUGI). It has been running an annual conference since 1976 and has about 3,500 attendees. For information about SUGI and other user groups, contact SAS Institute at 919-677-8000 or by using its World Wide Web universal resource location (URL):

```
http://www.sas.com
```

Courses

Most organizations offer "short courses" that cover different aspects of SAS use. These are usually free, informal, noncredit sessions. They are especially good for learning the site-specific details of running SAS. Contact your organization's help desk for the course schedule.

SAS Institute and a variety of third-party organizations offer training. These courses are sometimes expensive and may require travel to a training center. They are usually very high-quality and comprehensive, and may be customized to fit an organization's special needs. Contact the training group at SAS Institute for information on their courses — 919-677-8000.

The Institute also produces video courses and computer-based training modules. Consult your help desk staff to see which of these resources are available at your site.

Electronic Resources

If the person-to-person approach doesn't work for you, a variety of electronic means are at your disposal.

Online Help

Interactive environments, especially the SAS Display Manager (described in Appendix A), offer powerful online help. Get help by entering `help` at the SAS prompt or command line. Display Manager users may click on `help`, then click on `sas system` or `extended help` (these, and other choices, vary according to the operating system). Online help is especially useful when you need a quick memory jog about the syntax of a statement.

SAS/Assist

SAS Institute licenses numerous "add-on" products that enhance the functionality and ease of use of the SAS System. SAS/Assist is, as its name suggests, an interface that simplifies the use of many SAS procedures. It has a point-and-click interface that leads you through the steps necessary to carry out data management, reporting, and analysis tasks. Among its disadvantages: It can be tedious to use and it doesn't handle many DATA step tasks well.

Contact your help desk to see if SAS/Assist is available at your site. Or try the following: In Display Manager, click on Globals, then click on SAS/Assist. If it is installed on your computer, you will see a menu. If not, an error message appears in the SAS Log, indicating that SAS/Assist is not available.

SAS Sample Library

SAS Institute distributes a library of programs that demonstrate both basic and advanced features of its products. This "sample library" is a good source of ideas for all users, particularly novices. Ask your help desk for its location. Users running the SAS System, Version 6.10, under Windows can gain access to the library by clicking on `help`, then on `sample programs`.

RANK procedure, 205, *206*
STANDARD procedure, 203
TTEST procedure, 117
UNIVARIATE procedure, 104-105
VARCLUS procedure, 83
VARDEF option, MEANS procedure, 103
Variable lists, 65-66
Variables, 8. *See also* SAS datasets
data type, 18-19, 221, 293-294
describing. *See* LABEL statement
measurement issues, 8, 87, 93, 98, 103, 111, 155, 164, 167
VBAR statement, CHART procedure, 92, *93, 94*
VERIFY function, 225, 226-227
VIF option, REG procedure, 142
VW option, PROC NPAR1WAY, *165*

$w format, *45*
w. format, *24, 45*
w.d format, *24, 45*
Wald chi-square, 171
WEEKDATE format, *192*
WEEKDATX format, *192*
WEEKDAY format, *192*
WEEKDAY function, *191*
WEIGHT statement, 162, *163, 175*
WHERE statement, 61, 64-65, 289-290
used to eliminate outliers, 136
WILCOXON option, PROC NPAR1WAY, *165, 166*
Windowing environment. *See* Display Manager
Within-subjects designs, 130
WORDDATE format, *192*
WORDDATX format, *192*
World Wide Web, 278, 280
Writing observations. *See* OUTPUT statement
WSAVE command, Display Manager, 272, *279*

X command, Display Manager, *273*

YEAR function, *191*
YYMMDD format, *191*

Z command, Display Manager, *273*
Z format, *45*
Z scores, 202
Zellner's seemingly unrelated regressions, 82

The SAS-L List Server

Users with Internet access may subscribe free of charge to the SAS-L list server. This is an electronic, worldwide SAS discussion group with over 2,000 participants. Topics include routine problem solving, requests for information about competing or complementary products, job postings, and the occasional "flame war." SAS-L has a well-deserved reputation in cyberspace for being one of the most active, literate, and helpful discussion groups.

Joining SAS-L is simple. Send a message to

```
listserv@uga.cc.uga.edu
```

No subject is needed. The text of the message should read

```
sub sas-l your_name
```

If, for example, user Would B. Guru wanted to join, the message would be

```
sub sas-l Would B. Guru
```

Once the list server receives your request, it may forward it to one of the list's worldwide "peers." Once the appropriate server has processed your request, it notifies you that you have been added to the list. The notification message also includes information about how to post messages, review the posting database, and cancel your subscription.

World Wide Web

SAS Institute may be found on the World Wide Web. Point your Web browser (Netscape, Mosaic, and so on) to the uniform resource locator (URL)

```
http://www.sas.com
```

This resource is rapidly growing in popularity. It contains information about SAS Institute products, technical support, and other features of interest to both beginning and advanced SAS users.

Hard Copy

Many, if not most, people are still most comfortable seeking help and ideas from printed materials. As you will see in the following pages, there is no shortage of options from SAS Institute and other sources.

Local Documentation

Most organizations produce documents that explain how to run SAS at their sites. These documents are essential, since they deal with site-specific details such as queue names and printer names and locations. Ask your help desk personnel about these and other, non-SAS, publications.

SAS Institute Publications

SAS Institute produces a vast number of documents designed for users at all levels of expertise. For information on any of the publications described in this section, call SAS Institute at 919-677-8000, extension 7001. Note that some of these publications are also available at college and commercial bookstores.

Anyone interested in pursuing SAS programming at a level beyond this book should purchase

- SAS Language: Reference Version 6, first edition
- *SAS Procedures Guide Version 6,* third edition
- *SAS Software: Abridged Reference Version 6,* first edition

These are comprehensive, dictionary-style references. They cover DATA step features, the Display Manager, and the basic procedures. Users who want more task-oriented publications should consider

- *SAS Language and Procedures: Usage Version 6,* first edition
- *SAS Language and Procedures: Introduction Version 6,* first edition
- *SAS Applications Guide,* 1987 edition

These publications have many well-developed examples and focus on representative "real-world" programming and analysis tasks.

The *Companion* series of books describes operating system–specific features of SAS. *Companions* are available for all supported platforms. If you plan to do more advanced SAS programming, the *Companion* for your operating system is probably a good investment. Bear in mind, however, that you will still need site-specific documentation (described in the preceding section).

The Institute offers several non–manual format publications. *Observations* is a quarterly journal featuring meaty, in-depth articles. *SAS Communications* is a free quarterly magazine that has news, marketing, and technical tips. SAS User Group International (SUGI) conference proceedings may also be purchased. These are an excellent, eclectic collection of papers presented by leading SAS users as well as SAS Institute personnel. Finally, *SAS Views*, the bound version of the Institute's class notes, is for sale.

Books by Users

Non-Institute authors offer different perspectives on SAS software. SAS Institute offers over 25 such publications as part of its "Books by Users" program. Of particular interest to SAS novices:

- *Mastering the SAS System,* second edition, Jay Jaffe
- *SAS Applications Programming: A Gentle Introduction,* Frank DiIorio
- *SAS Software Solutions: Basic Data Processing,* Thomas Miron
- *SAS Programming for Researchers and Social Scientists,* Paul Spector
- *Working with the SAS System,* Eric W. Tilanus

Common Problems (and Solutions)

Despite your best intentions, it's almost inevitable that you will make syntax and/or logic errors in your programs. Even seasoned professional programmers do this, so it should not alarm beginners. What *is* important is being able to recognize situations that cause problems and to know how to prevent them in the future. In this section, we present some of the more common syntactical and logical mistakes made by new users of the SAS System. Keep in mind that the Logs being displayed were generated by Version 6.10 of SAS running under Windows. Logs from other versions and platforms may appear somewhat different, but the spirit, and certainly the solutions we offer, are the same for all environments.

Division by Zero

The Problem: Even the simplest division calculations can cause SAS to complain if the denominator is inappropriate. Exhibit C.1 shows a Log in which a value for the denominator (variable `miles`) had a value of 0 in an observation. SAS prints a message identifying the offending calculation (stating the obvious here, but very useful in long programs). Then it prints a ruler and the offending line of raw data, followed by a dump of all values known to the DATA step for the observation. A final note is displayed to tell you that missing values were created because the mathematical operation could not be carried out.

Exhibit C.1

Log Indicating Division by Zero

```
1     DATA _null_;
2     INPUT outlay miles;
3     per_mile = outlay / miles;
4     CARDS;

NOTE: Division by zero detected at line 3 column 19.
RULE:----+----1----+----2----+----3----+----4----+----5----+
7     140 0
OUTLAY=140 MILES=0 PER_MILE=. _ERROR_=1 _N_=3
NOTE: Mathematical operations could not be performed at the
      following places.
      The results of the operations have been set to missing
      values. Each place is given by: (Number of times) at
      (Line):(Column).
      1 at 3:19
```

A Solution: It is not necessarily wrong to leave your program the way it is in Exhibit C.1. You would probably want the calculated variable (`per_mile`) to take the default, or missing, value. A cleaner way to program the calculation is to bypass it if the denominator is not suitable, as follows:

```
IF miles ^= 0 THEN per_mile = outlay/miles;
```

`per_mile` will not be calculated when `miles` is 0, and thus will be set to missing.

Unbalanced Quotes

The Problem: While using Display Manger, you submit a short program, followed by a RUN statement. Nothing happens. Control is immediately returned to the Program Editor window. You issue a RECALL command and resubmit the program. The result is shown in Exhibit C.2.

Exhibit C.2

Unbalanced Quotes

```
1     DATA baseline;        /* Begin First SUBMITted program */
2     INFILE '\survey\rawdata\panel1.dat';
3     INPUT co_id year qtr netinc;
4     IF revenue > 0 THEN type = 'profit';
5        ELSE             type = 'loss;
6     RUN;
7
8     PROC PRINT DATA=baseline;
9     RUN;
10    DATA baseline; ;      /* Begin Second SUBMITted program */
11    INFILE '\survey\rawdata\panel1.dat';
12    INPUT co_id year qtr netinc;
13    IF revenue > 0 THEN type = 'profit';
                                 ------
                                 388
14       else             type = 'loss;
15    RUN;
```

Exhibit C.2

Unbalanced Quotes (continued)

```
ERROR 388-185: Expecting an arithmetic operator.

NOTE: The SAS System stopped processing this step because of
      errors.
WARNING: The data set WORK.BASELINE may be incomplete. When
         this step was stopped there were 0 observations and
         6 variables.
NOTE: The DATA statement used 6.96 seconds.


16
17    PROC PRINT DATA=baseline;
18    RUN;

NOTE: No observations in data set WORK.BASELINE.
NOTE: The PROCEDURE PRINT used 0.5 seconds.
```

A Solution: Quoted strings, such as character constants, titles, and footnotes, need to be enclosed in quotation marks or apostrophes ("xx" or 'xx'). Syntax problems associated with this are forgetting to close the quote ("xx), and using different characters to do the quoting ("xx' or 'xx"). The result of these problems may also create an error message resembling this one:

```
ERROR: Quoted string exceeds 200 characters
```

A close look at line 5 reveals a quote that was not closed (the character constant `loss` in line 5). One solution is to SUBMIT the appropriate character — in this case, a single quote ('). This will complete the character constant and, most likely, get SAS's syntax checker back on track.

```
"; RUN;
```

Finally, you would RECALL the program, correct the imbalance, and resubmit the program.

DOing Without ENDing

The Problem: A DO group requires a DO statement and an associated END statement. If SAS does not find an equal number of DOs and ENDs in a DATA step, it will issue an error message similar to that shown in Exhibit C.3.

Exhibit C.3

DO-END Imbalance

```
18    DATA trans;
19    SET trial;
20    IF race = 1 THEN DO;
21       cost = cost * .9;
22       OUTPUT;
23    RUN;

23    RUN;
         -
         117
ERROR 117-185: There were 1 unclosed DO blocks.
NOTE: The SAS System stopped processing this step because of
       errors.
WARNING: The data set WORK.TRANS may be incomplete. When this
          step was stopped
          there were 0 observations and 4 variables.
```

A Solution: Insert an END statement. In this case, it would follow the OUTPUT statement, as follows:

```
IF race = 1 THEN DO;
   cost = cost * .9;
   OUTPUT;
   END;
```

The correct location of a missing END may be a bit harder to determine in long programs or in those that have multiple and nested DO groups. Indenting statements within DO groups is an effective and simple way to highlight the structure of the program and often prevents omission of the END statement.

Uninitialized Variables

The Problem: Variables must be assigned values *before* they are used in a calculation. This may be done by using an earlier calculation or INPUT, SET, or MERGE statements, which read variables into the DATA step. A variable is uninitialized if the first reference to it in the DATA step forces SAS to say, in effect, "I haven't heard of this variable, but here you are using it in a calculation. I'll set it to missing [numeric variable] or blank [variable]." Exhibit C.4 is just one scenario that will generate the notes shown in the Log.

Exhibit C.4

Uninitialized Variables

```
40    DATA testv2;
41    SET test;
42    IF faminc > 50000 THEN group = h;
43       ELSE                group = l;
44    RUN;

NOTE: Variable H is uninitialized.
NOTE: Variable L is uninitialized.
NOTE: The data set WORK.TESTV2 has 1 observations and 9
      variables.
```

Some Solutions: We inadvertently forgot to quote the values `h` and `l`. Rather than assigning *constant* values to GROUP, SAS was instructed to assign *variable* values. The variables `h` and `l` could not be located. This generated the "uninitialized" messages in the Log. The solution in this example is as follows:

```
IF faminc > 50000 THEN group = 'h';
   ELSE                group = 'l';
```

Another reason for generating this message is the misspelling of a variable name. In Exhibit C.4, you might have specified variable `h` instead of `high`, or `l` instead of `low`.

Subtle Omissions of Periods

The Problem: Both system and user-written formats must end with a period (.) to allow SAS to distinguish between variable and format names. Exhibit C.5 shows the sometimes puzzling message that is produced when you inadvertently omit a period from a format name.

Exhibit C.5

Missing Period in a Format Specification

```
28    DATA test;
29    INPUT id faminc panel1a panel1b panel2a panel2b;
30    FORMAT faminc dollar8;
31    CARDS;

NOTE: Variable DOLLAR8 is uninitialized.
NOTE: The data set WORK.TEST has 1 observations and 6 variables.
```

A Solution: Once you recognize that the uninitialized variable is actually the name of a format, the solution is easy. Simply add a period, as follows:

```
FORMAT faminc dollar8.;
```

Display Manager "Stalls"

The Problem: *Everyone* makes this mistake now and then. Working in Display Manager, you use the SUBMIT command to process one or more statements. The statements immediately appear in the Log window — and do nothing. The SAS System immediately returns control to the Program Editor window. You wait (and wait) for something to happen.

The Solution: You forgot to tell SAS that you actually want the program to run. The only way interactive programs are actually executed is by entering a RUN statement, as follows:

```
RUN;
```

This omission has inspired untold numbers of caustic and, ultimately, embarrassing comments about system response time. It's an honest mistake, and easily fixed.

Data Type Conversion Messages

The Problem: Some types of data are clearly numeric (salary, age); others, clearly character (name, address, title). Some data, however, fall into a gray area and could be coded either way. In Exhibit C.6, questionnaire responses (q1 and q2) are read as character data (line 46). In the next DATA step, however, we use them in a *numeric* function (line 55). The SAS System prints a note in the Log saying that it converted character values to numeric ones.

Exhibit C.6

Character to Numeric Conversion

```
45    DATA baseline;
46    INPUT id year $ q1 $ q2 $;
47    CARDS;

NOTE: The data set WORK.BASELINE has 4 observations and 4
      variables.
NOTE: The DATA statement used 0.48 seconds.


52    RUN;
53    DATA alter;
54    SET baseline;
55    avg = MEAN(q1, q2);
56    RUN;

NOTE: Character values have been converted to numeric
      values at the places given by: (Line):(Column).
      55:12   55:16
NOTE: The data set WORK.ALTER has 4 observations and 5
      variables.
```

A Solution: This message is a signal that you did not correctly recall the data types in your dataset. In some cases the conversion will be done correctly. A character "5" will be treated as the number 5, for example. You should not rely on the accuracy or the consistency of SAS's automatic conversion. Instead, the data type should be changed by rewriting the INPUT statement. The following INPUT statement will read the variables correctly and avoid the conversion note later on:

```
INPUT id year $ q1 q2;
```

New Character Variables Appear Truncated

The Problem: A series of IF-THEN-ELSE statements assigns values to a character variable. When you use the variable, some of the characters on the right have been lost. Such a program appears in Exhibit C.7. Variable `incgroup` was not already in dataset `master95`.

Exhibit C.7

Truncated Character Variables

```
LIBNAME sasdata 'user:[fcd.book2]';
DATA modified;
SET sasdata.master95;
IF            . <= income <  30000 THEN incgrp = 'Low';
   ELSE IF 30000 <= income <= 70000 THEN incgrp = 'Middle';
   ELSE                                  incgrp = 'High';
```

A Solution: SAS uses the length of the first reference to the character variable `incgrp`. Since this reference was to assign the value 'Low', `incgrp` is created with a length of 3. The longer values of 'Middle' and 'High' are stored in three bytes as well, and lose the fourth through last characters. No warning message about this truncation is printed in the Log! A solution is to use the LENGTH statement, discussed in Chapter 14. It should precede the first reference to `incgrp`, as shown here:

```
SET sasdata.master95;
LENGTH incgrp $3;
IF     . <= income <  30000 THEN incgrp = 'Low';
```

Subsetting into 0 Observations

The Problem: A WHERE statement or an IF-THEN-OUTPUT statement created a dataset with 0 observations.

Exhibit C.8

WHERE Reduces to 0 Observations

```
69   DATA baseline;
70   INPUT id montth q1 $ q2 $;
71   CARDS;

NOTE: The data set WORK.BASELINE has 4 observations and 4
      variables.
NOTE: The DATA statement used 0.6 seconds.

76   RUN;
77
78   PROC PRINT DATA=baseline;
79   WHERE month = "JU";
80   RUN;

NOTE: No observations were selected from data set
      WORK.BASELINE.
```

Some Solutions: There are a few possibilities here. There may really not be any months equal to "JU", so a result of 0 observations is reasonable, if disappointing. Check to see that month is really coded as the first two characters of the month. It may instead be "01", "02", and so on. Finally, see if the dataset is supposed to span the time period that includes June.

A more complicated situation that can also generate this message occurs when the WHERE statement's logical condition is impossible. For example,

```
WHERE gender = '1' & pregnant = 'Y';
```

If the gender coding for females was actually "2" instead of "1", WHERE would clearly be at a loss, since you would be looking for pregnant males. Probably the only good to come out of this WHERE statement is that it identifies observations with erroneous coding of gender.

Complaints About Correct Syntax

The Problem: SAS is issuing mysterious error messages, complaining about program statements that appear valid. The SAS System begins its negative critique of your program on line 87 of Exhibit C.9. The statement is valid, yet it and other seemingly correct statements are flagged as incorrect.

Exhibit C.9

Mysterious Error Messages

```
81    * \thesis\sasprogs\baseline.sas
82
83      Reads baseline data, performs transformations
84      as discussed in 11/19 meeting with committee.
85
86    DATA baseline;
87    INPUT id year $ q1 $ q2 $;
      -----
      180

ERROR 180-322: Statement is not valid or it is used out of
               proper order.

88    CARDS;
      -----
      180

ERROR 180-322: Statement is not valid or it is used out of
               proper order.

89    154 FR 3 1
      ---
      180
90    110 FR 4 4
91    115 SO 8 7
92    100 SO 8 8
93    RUN;
ERROR 180-322: Statement is not valid or it is used out of
               proper order.
```

The Solution: Error 180, `Statement is not valid`, is often your clue that either SAS or you lost track of where one statement ended and another began. It is almost always your doing, and almost always the result of a missing semicolon. Look at the end of line 84. It is the end of the comment that began in line 81, but does not have a semicolon. The comment will not end until SAS encounters a semicolon. The semicolon is finally located at the end of line 86. Although it *appears* that the DATA statement was present, from SAS's perspective it was simply part of the comment. The first statement SAS tries to execute is INPUT. The SAS System is rightly confused. Adding a semicolon to the comment, as shown below, will correct the problem.

```
as discussed in 11/19 meeting with committee.;
```

Unbalanced Parentheses

The Problem: A seemingly correct calculation is flagged because of a missing right parenthesis. The calculation appears to be written correctly.

Exhibit C.10

Unbalanced Parentheses

```
1   DATA tranform;
2   INPUT id score1 score1a score2 score2a;
3   tot = (MAX(score1, score1a) + MAX(score2, score2a) / 2;
                                                         -
                                                         73
4   CARDS;

ERROR 73-322: Expecting an ).

NOTE: The SAS System stopped processing this step because of
      errors.
WARNING: The data set WORK.TRANFORM may be incomplete. When
         this step was stopped there were 0 observations and 6
         variables.
WARNING: Data set WORK.TRANFORM was not replaced because this
         step was stopped.
```

A Solution: The numbers of left and right parentheses in a statement must always be equal. Upon closer inspection of line 3, it is clear that there are more left parentheses (3) than right (2). A right paren should be inserted before the division sign. Even in relatively simple statements such as this one, it's helpful to use multiple lines and blank space to clarify the calculation and preempt syntax errors. The following rewrite of line 3 is easier to follow:

```
tot = (MAX(score1, score1a) +
       MAX(score2, score2a))
       / 2
```

"Dataset Not Found" Messages

The Problem: The SAS System is unable to locate a dataset used in a DATA step or a PROC. Exhibit C.11 shows one of the error messages produced.

Exhibit C.11

"Dataset Not Found"

```
1 DATA mods;
2 SET sasdata.fy1995;
ERROR: Libname SASDATA is not assigned.
3 IF year = 1995 THEN DO;
4     yprime = y * scale;
5     OUTPUT;
6     END;
7 RUN;

NOTE: The SAS System stopped processing this step because of
      errors.
WARNING: The data set WORK.MODS may be incomplete. When this
         step was stopped there were 0 observations and 4
         variables.
```

Some Solutions: In this case, the two-level dataset name `sasdata.fy1995` could not be located because the LIBNAME for `sasdata` had not been issued before the DATA statement. A LIBNAME resembling the following should precede the DATA step:

```
LIBNAME sasdata "path";
```

In the above, *path* identifies the directory or system dataset in which the `fy1995` dataset is located. If a LIBNAME had been issued and the SAS System could not locate the dataset, it might be due to a misspelling — in Display Manager, use the DIR window to review dataset names. In batch mode, run the following procedure:

```
PROC CONTENTS DATA=sasdata._ALL_;
RUN;
```

Every Calculated Value Is Missing

The Problem: The Log reports that a calculation resulted in a missing value for each observation in a dataset. Exhibit C.12 shows a few clues to this puzzle.

Exhibit C.12

All Calculated Values Are Missing

```
1     DATA milestat;
2     INPUT outlay 1-4 miles 5-10;
3     per_mile = outlay / mileage;
4     CARDS;
NOTE: Variable MILEAGE is uninitialized.
NOTE: Mathematical operations could not be performed at the
      following places.
      The results of the operations have been set to missing
      values. Each place is given by: (Number of times) at
      (Line):(Column).
      40 at 3:19
NOTE: The data set WORK.MILESTAT has 40 observations and
      4 variables.
```

A Solution: Exhibit C.12 reveals that there were 40 instances of missing values being created during a calculation. There were also 40 observations in the dataset. This is your clue that something was uniformly wrong with the data or with the calculation. The note that the variable `mileage` is uninitialized is important. Given the INPUT statement's variables, it's likely that the calculation in line 3 should have used `miles`. The misspelling ensures that every observation will have a missing value for `per_mile`.

This situation also arises when the INPUT statement reads the data incorrectly. If, for example, `miles` were actually in columns 5 through 8 (rather than 5 through 10), it is possible that all the data would be read in blank columns and thus would be stored as missing values. The solution in this case is to use the PRINT procedure to display the data and make corrections if the values are not what you expected.

Conflicting Data Types

The Problem: SAS issues an error message that a variable has been "defined as both character and numeric." This is shown in Exhibit C.13.

A Solution: When two or more datasets are being combined using a SET or MERGE statement, like-named variables must be of the same data type (character or numeric). In Exhibit C.13, variables `q1` and `q2` were character in dataset `baseline` (line 116), but numeric in dataset `round1` (line 125). One of the INPUT statements should be changed to make the data types consistent.

Exhibit C.13

Conflicting Data Types When Combing Datasets

```
115  DATA baseline;
116  INPUT id year $ q1 $ q2 $;
117  CARDS;
NOTE: The data set WORK.BASELINE has 4 observations and 4
      variables.
NOTE: The DATA statement used 0.6 seconds.
122  RUN;
123
124  DATA round1;
125  INPUT id year $ q1 q2 ;
126  CARDS;
NOTE: The data set WORK.ROUND1 has 4 observations and 4
      variables.
NOTE: The DATA statement used 0.5 seconds.
131  RUN;
132
133  DATA combine;
134  SET baseline round1;
ERROR: Variable Q1 has been defined as both character and
       numeric.
ERROR: Variable Q2 has been defined as both character and
       numeric.
135  RUN;
NOTE: The SAS System stopped processing this step because of
      errors.
WARNING: The data set WORK.COMBINE may be incomplete. When
         this step was stopped there were 0 observations and
         4 variables.
```

All Calculations Are Missing

The Problem: Nonmissing values are entered for variables used in a calculation. The PRINT procedure indicates that you read the data correctly. When the time comes to use the calculated variable in a PROC, however, you see that all occurrences of the variable are missing. When you created the dataset, there were no messages such as "Mathematical operations could not be performed" or "Results of the operations have been set to missing." The Log is shown in Exhibit C.14.

Exhibit C.14

Statements Are Sequenced Incorrectly

```
1     DATA health;
2     INPUT id $ sex $ weight h_ft h_inch;
3     IF sex = 'M' THEN DO;
4        OUTPUT;
5        h_cm = 2.52 * ((12 * h_ft) + h_inch);
6        END;
7     CARDS;

NOTE: The data set WORK.HEALTH has 10 observations and 6
      variables.
NOTE: The DATA statement used 2.14 seconds.

19    RUN;
```

A Solution: In programming, as in comedy, timing is everything. You want to calculate and output only for those observations where variable `sex` equals 'M'. `h_cm`, in keeping with SAS System rules, is set to missing at the top of each pass through the DATA step, and is assigned a value at line 5. The observation, however, is written to dataset `health` in line 4. This ensures that `h_cm` will always be missing when the observation is output. Reversing the order of these statements will correct the missing-value problem. This sort of problem is particularly nasty because there is no *syntactical* error. Rather, it is a *logical* error, which does not necessarily appear in the Log with a note, warning, or error message.

Index

Note: Italicized numbers indicate Exhibit references.

? command, Display Manager, 272
| (OR) operator, *38*
|| concatenation operator, 223-224
& (AND) operator, *38*
/ line pointer, *23*
: comparison modifier, 224

A command, Display Manager, *274, 275*
ALL reserved dataset name, 74
Akaike Information Criterion, 171
ALPHA option
    CORR procedure, 82
    MEANS procedure, 103
Analysis of variance, 111, 112. *See also* ANOVA procedure; GLM procedure
Analysis of covariance, *See* GLM procedure, for ANCOVA; REG procedure, for ANCOVA
AND logical operator, *39*
ANOVA procedure, 80
Ascending option, CHART procedure, 96
Assignment statement, 35-37, 289. *See also* Functions
    compared to user-written formats, 255
    examples, 36-37, *160*
    impact of missing values, 36
    invalid operations, 283-284
    syntax, 35
usage rules, 35-36, 222
AUTOREG procedure, 81, 142
Autoregressive processes, 81

B command, Display Manager, *274*
Balanced designs, 127
Bar charts. *See* CHART procedure
Batch processing, 10
Beta coefficients, 80. *See also* Path models
BON option, GLM procedure, *121, 122*
Bonferonni t test, *121*
Books By Users program, 282
BOTTOM command, Display Manager, 268, *272*
Box plots, 104, 107, 109
    side-by-side, 115, *116*
Brown-Mood K-group test, *159*
BY statement
    used to combine SAS datasets, 47, 53, *54*, 55, *56*
    used with PRINT procedure, *70*
    used with procedures, 61, 62
    used with SORT procedure, 72-73
    used with UNIVARIATE procedure, 115
BYE command, Display Manager, 270

C command, Display Manager, *274, 275*
Calculations. *See* Assignment statement; Functions
CALIS procedure, 82
CANCORR procedure, 82
CANDISC procedure, 83
Canonical correlation analysis, 82
CARDS statement, 17
    syntax, 20
CATMOD procedure, 81, 155, 174-179
    compared with LOGISTIC procedure, 167
CEIL function, *188*

CENTER system option, *186*
Character data, 221-229, 285, 286-287. *See* also Variables, data type
  functions, 224-229
  properties, 221-222, 289
CHART procedure, 91-101, 111
CHISQ option, FREQ procedure, 159, *160*
CLASS statement
  GLM procedure, 118, *119, 125*
  MEANS procedure, *196*, 197
  NPAR1WAY procedure, 165, *166*
  TTEST procedure, 117
CLEAR command, Display Manager, *272*
CLM option, MEANS procedure, 102, *103*
CLUSTER procedure, 83
Cohen's kappa, 79
Collinearity, 142
COLLINOINT option, REG procedure, 142
Column input data, 22-23. *See also* INPUT statement
Combining SAS datasets, 47-59
  concatenation, 48, 50-53
  impact of missing values, 49
  interleaving, 48, 53-54
  matched merge, 48, 54-57
  methods, 48
  scenarios for, 47
  usage notes, 48-49
COMMA format, *24, 44*
Comments, 32-33, 290-291
Comparison operators, 37-38, 40, 223-224
COMPBBL function, 225, 228
COMPRESS function, 225, 228
Concatenating datasets. *See* Combining SAS datasets, concatenation
Confidence interval, 103, 140
Confirmatory factor analysis, 82
CONTENTS procedure, 74-75. *See also* SAS datasets
CONTRAST statement, GLM procedure, 121-122
CORR procedure, 79, 82
CORRESP procedure, 82
Correlation. *See* CORR procedure
Cramer's V, 79
Cramer-von Mises K-group test, *159*
Cronbach's alpha, 82
Crossed effects. *See* MODEL statement, GLM procedure
Crosstabulations. *See* FREQ procedure
Custom formats. *See* FORMAT procedure
CV option, MEANS procedure, *198*

D command, Display Manager, *274, 275*
DATA statement, 17
  syntax, 19
DATA step, 8-9, 31, 183
  alternatives, 195
Data type. *See* Variables, data type
DATALINES statement, 20
Date constants, 190
DATE format, *191*
DATE system option, *186*
Date-oriented data, 189-192
  formats, *191*
  functions, *191*
DAY function, *191*
DDMMYY format, *191*
Default settings, 10
DESCENDING option
  CHART procedure, 96
  LOGISTIC procedure, 169, *172*, 174
Descriptive statistics. *See* MEANS procedure; UNIVARIATE procedure
Differences of related means, 128, *129*
DISCRETE option, CHART procedure, 93, *94*, 98
DISCRIM procedure, 83
Display Manager, 11, 184, 261-275, 284-285, 287-288. *See also* Interactive processing
  advantages, 261
  command syntax, 265
  examining output, 268
  invoking, 262
  issuing commands, 263-265, 269
  screen layout, *262, 263*, 272
  writing programs using, 266, 268-269
Displaying data. *See* PRINT procedure
Distributed lag model, 81
DO groups, 187, 285-286
DO statement, 187. *See also* END statement
Documenting programs. *See* Comments
DOLLAR format, *44*
DOWN command, Display Manager, 268, *272*
DROP option, SET statement, 33
Duncan multiple range test, *121*
Durban-Watson statistic, 141-142
Dummy variables, 148, 149, 179, *240-241*, 242, *243-244*
DUNCAN option, GLM procedure, *121*
DW option, REG procedure, 141-142

E format, *44*
EDF option, PROC NPAR1WAY, *165, 166*
END statement, 187. *See also* DO statement
EQ comparison operator, *38*
EQS notation, 82
Errors, 11, 44
ERRORS system option, *186*
EXACT option, FREQ procedure, 159, *160*
Execution. *See* Program execution
Exiting SAS, 270
Exploratory factor analysis, 82

FACTOR procedure, 82
Factorial design, 124
FASTCLUS procedure, 83
FILE command, Display Manager, 269-270, 272

FIRSTOBS system option, *186*
Fisher's exact chi-square, 79, 159
Fixed effects models, 79
FLOOR function, *188*
FMTERR system option, *186*
FOOTNOTE statement, 61, 62-63
FORMAT procedure, 209-219
    concepts, 210
    syntax, 210-212
FORMAT statement
    used in DATA step, 43-45
    used with PROCs, 61, 63-64, *70, 212, 213, 214, 215, 216*
Formats, 23-24, 44-45. *See also* Date-oriented data, formats; FORMAT procedure
    problems using, 25-26, 287
    usage notes, 44
Formatted input data, 23-24. *See also* INPUT statement
FRACTION option, MEANS procedure, 204-205
FREQ option, UNIVARIATE procedure, 107
FREQ procedure, 78, 79, 87-91
    crosstabulations, 155, 156-164
Frequency distributions. *See* FREQ procedure
Function keys. *See* KEYS command, Display Manager; KEYS window
Functions, 36, 183, 187-189, 288. *See also* Character data, functions; components, 187-188; Date-oriented data, functions

Gamma, 79
GE comparison operator, *38*
GLM procedure, 80, 111
    for ANCOVA, 143-151
    for one-way ANOVA, 118-124
    for two- to n-way ANOVA, 124-132
Goodman and Kruskal's Gamma, 171
GROUP option, CHART procedure, 113, *114*
GROUPS option, RANK procedure, 204-205, *206*
Grouping values. *See* Recoding
GT comparison operator, *38*

HBAR statement, CHART procedure, 92
HELP command, Display Manager, 272, 279
Help desks, 277-278
Histograms, 98, *99*
Hosmer-Lemeshow goodness-of-fit test, 170

ID statement
    PRINT procedure, 68, *70*
    UNIVARIATE procedure, 105, 107
IF-THEN-ELSE statements, 39-41
    alternatives to, 209
    examples, 40-41
    syntax, 40
IN
    comparison operator, *38, 39*
    dataset option, 50, 55, *57, 59*
INCLUDE command, Display Manager, 270-271, *272*
Incomplete data. *See* Missing data
INDEX function, 225, 226, 227
INDEXC function, 225, 226, 227, 229
INDEXW function, 225, 226
INFILE statement, 17
    example, *20*, 21
    syntax, 20
INFLUENCE option, LOGISTIC procedure, 170
Information centers. *See* Help desks
INPUT statement, 17
    problems reading data, 25-26
    styles, 21, 24-25, *175*
    syntax, 21
Interactions, 148, *149*. *See also* MODEL statement, GLM procedure
Interactive processing, 10. *See also* Display Manager
Interleaving datasets. *See* Combining SAS datasets, interleaving
Item-to-total correlations, 82

KEEP option, SET statement 33, *34, 35*
Kendall's tau, 79, 171
KEYS command, Display Manager, 271, *273*
KEYS window, *271*
Kolmogorov test statistic, 107
Kolmogorov-Smirnov test of differences, *159*
Kruskal-Wallis test, 159
Kuiper two-group test, *159*
KURTOSIS option, MEANS procedure, *198*

LABEL statement
    used in DATA step, 43
    used with PROCs, 61, 62-63
LACKFIT option, LOGISTIC procedure, 170, *172*
Lambda, 79
Latent variables, 82
LE comparison operator, *38*
Least significant difference pairwise t tests, *121*
LEFT command, Display Manager, 268, *273*
LEFT function, 225
LENGTH function, 225
LENGTH statement, 222-223, 289
LEVELS option, CHART procedure, 100, *101*
LIBNAME statement, 18, 27-28, 291-292. *See also* SAS datasets
LIFEREG procedure, 81
LIFETEST procedure, 81
Likelihood ratio, 79
Linearity assumptions, 133, 140
LINESIZE system option, *186*
LINEQS notation, 82
LISREL models, 82
List input data, 21-22. *See also* INPUT statement
Listwise deletion, 139
LOG command, Display Manager, 269
LOG function, *188*

LOG10 function, *188*
Logical expressions, 37-39
  comparison operators, 37-38
  logical operators, 38-39
LOGISTIC procedure, 81, 155, 167-174
  compared with CATMOD procedure, 167
Logistic regression, 81. *See also* CATMOD procedure; LOGISTIC procedure
LOWCASE function, 225, 228-229
LSD option, GLM procedure, *121*, *122*
LT comparison operator, *38*

M command, Display Manager, *274*
Mann-Whitney U two-group test, *165*
MAX function, *189*
MAXDEC option, MEANS procedure, 102, *103*
MDS procedure, 82
MEAN function, *188*
MEAN option
  MEANS procedure, 102, *103*
    in OUTPUT statement, *198*, 199
  STANDARD procedure, 202, *203*
MEANS procedure, 78, 98, 102-104
  compared to UNIVARIATE procedure, 102
  used to create output datasets, 195-203
MEANS statement, GLM procedure, 118, *119*, *125*
Measurement. *See* Variables, measurement issues
MEASURES option, FREQ procedure, 159, *160*
MEDIAN option, NPAR1WAY procedure, *159*
MERGE statement, 47, 54-55. *See also* Combining SAS datasets, matched merge
  examples, 55, *56-57*, *58-59*
  usage notes, 55
Merging datasets. *See* Combining SAS datasets, matched merge
MIDPOINTS option, CHART procedure, 98-100
MAX option, MEANS procedure, 102, *103*
  in OUTPUT statement, *198*
MIN function, *189*
MIN option, MEANS procedure, 102, *103*
  in OUTPUT statement, *198*
Missing data, 8, 19
  creating from nonmissing data, 234, *235*
  handling when combining datasets, 49
  impact on calculated variables, 36, 188, *246*
  impact on comparison operators, 38
  impact on logical comparisons, 38
  in CHART procedure, 96
  in FREQ procedure, 90
  in LOGISTIC procedure, 170
  in REG procedure, 139
  in TTEST procedure, 118
MISSING option
  CHART procedure, 96
  FREQ procedure, 90
  MEANS procedure, 196-197
MISSING system option, *186*
MIXED procedure, 80
MMDDYY format, *191*
MODECLUS procedure, 83
MODEL statement
  CATMOD procedure, 174, *175*
  GLM procedure, 118, *119*, *125*
    specifying interactions or crossed effects, 126-127, 144
  LOGISTIC procedure, *168*, 179
  REG procedure, 136, *137*
MODEL procedure, 82
Modes of execution, 10, 261
MONTH function, *191*
MONYY format, *191*
Multicollinearity, 80, 142
Multidimensional scaling, 82
Multinomial logistic regression models, 81
Multiple comparison methods, 121
Multiple R squared, 80

N function, *189*
N option
  MEANS procedure, 102, *103*
    OUTPUT statement, *198*
  PRINT procedure, 69, *70*
NEGPAREN format, *44*
Nested DO groups, 187
NESTED procedure, 80
NLIN procedure, 80
NMISS function, *189*
NMISS option, MEANS procedure, *198*
NOCOL option, FREQ procedure, 159, *160*
NOCUM option, FREQ procedure, 90, *91*
NODUPKEY option, SORT procedure, 73
NODUPREC option, SORT procedure, 73
Nonlinear regression, 80
NOPERCENT option, FREQ procedure, 159, *160*
NOPRINT option
  MEANS procedure, 196-197
  STANDARD procedure, 203
NORMAL option, UNIVARIATE procedure, 107, *108*, 109
NOROW option, FREQ procedure, 159, *160*
Normal probability plots, 104, 107
NOSTATS option, CHART procedure, 96
NOTIN comparison operator, *38*
NOU option, GLM procedure, 131
NOUNI option, GLM procedure, 130, *131*
NMISS option, MEANS procedure, 102
NPAR1WAY procedure, 155, 164-167
NUM command, Display Manager, *273*
NUMBER system option, *186*
Numeric variables. *See* Variables, data type
NWAY option, MEANS procedure, 196-197

OBS dataset option, 33, *34*, 68
OBS system option, *186*
Observations, 8. *See also* SAS datasets
Operators. *See* logical expressions
Options, 184-186. *See also* OPTIONS procedure
    commonly used, *186*
    identifying, *184-185*
    specifying, 185-186
OPTIONS procedure, 185
OPTIONS statement, 185
OPTIONS window, Display Manager, *273*
OR logical operator, *39*
ORDER option, FREQ procedure, 90, *91*
ORTHOREG procedure, 80
OUT option
    RANK procedure, 204
    SORT procedure, 72
Outliers, 133
    identifying with PLOT procedure, 136
    identifying with PLOT statement, REG procedure, 140-141
    using WHERE statement to eliminate, 136
Output file, *15*
OUTPUT command, Display Manager, 269
OUTPUT statement
    DATA step, 42-43, *65*, *232*
    MEANS procedure, *196*, 198-200

PAGENO system option, *186*
PAGES system option, *186*
PAGESIZE system option, *186*
Panel study, 81
Partial correlation. *See* CORR procedure
Partial sums of squares. *See* Sums of squares, type III
Path models, 80, 82, 136. *See also* CALIS procedure
PDLREG procedure, 81
Pearson chi-square, 174
Pearson r, 79
PERCENT format, *44*
PERCENT option, RANK procedure, 204-205, *206*
Phi, 79
PHREG procedure, 81
Planned tests, 112, 121
PLOT option, UNIVARIATE procedure, 107, *108*
PLOT procedure, 111, 133-136
PLOT statement
    PLOT procedure, 133, *134*
    REG procedure, 140-141
PMENU command, Display Manager, *272*
PMENUs. *See* Display Manager, issuing commands
Poisson regression models, 81
Population versus sample statistics, 103
Post hoc tests, 112, 121
PREDICTED diagnostic variable, REG procedure, 141
Previously crosstabulated data, 162, 174
PRINCOMP procedure, 82
PRINT procedure, 66-71
    syntax, 67-69
Probit models, 81
Procedures, 9
    caveats surrounding usage, 78
    naming conventions, 77-78
    overlapping functionality, 78
Programming style, iii, 9, 284, 286
Programs. *See* SAS programs
PROGRAM command, Display Manager, 268
Program execution, 10
PRT option, MEANS procedure, 102, *103*
PUT function, 216-217, *218*

R command, Display Manager, *274*
Random effects models, 80
RANGE option, MEANS procedure, *198*
RANK procedure, 195, 204-207
RANK statement, RANK procedure, 205
RANNOR function, *189*
RANUNI function, *189*
Raw data, 7, *12*
    elements, 17
    layout, *12*, 18
    reading. *See* INPUT statement
Rearranging SAS datasets. *See* SORT procedure
RECALL command, Display Manager, 268, *273*
Recoding, 41-42, 98, 209, 214-215, 216-217
REG procedure, 80, 82, 111, 136-143
    for ANCOVA, 148-151
    interactive use, 142-143
Regression analysis, 133. *See also* GLM procedure; REG procedure
Regression sums of squares. *See* Sums of squares, type III
Reliability. *See* ALPHA option, CORR procedure
RENAME dataset option, 50, *52*, *53*
Repeated measures, 80, 130
REPEATED statement, GLM procedure, 80, 130-132
REPLACE option, STANDARD procedure, 203
RESET command, Display Manager, *272*
RESIDUAL diagnostic variable, REG procedure, 141
Residual plots, 80, 140
RIGHT command, Display Manager, 268, *273*
RIGHT function, 225
ROUND function, *189*
RUN command, Display Manager, 267

Sample programs, 279
SAS courses, 278
SAS datasets, 7-8. *See also* CONTENTS procedure; LIBNAME statement
    contents, 26-27
    permanent versus temporary, 26
SAS Log, 11, *14*
SAS programs *See also* DATA step; Procedures
    structure, 7, 8-9, 12-13
SAS System

as integrated software, 1
competitors, 1
documentation, 83-85
scope, I, 2, 5
SAS publications, 281-282
SAS System options. *See* Options
SAS-L list server, 280
SAS/ASSIST, 279
saturated model, 178
SAVAGE option, NPAR1WAY procedure, *159*
Savage scoring of ranks, *159*
Scale. *See* Variables, measurement issues
Scatterplots. *See* PLOT procedure
Scheffe multiple comparison technique, *121*
Scheffe option, GLM procedure, 121
Selecting observations. *See* OUTPUT statement; WHERE statement
SET statement, 33-34, 47. *See also* Combining SAS datasets, concatenation; Combining SAS datasets, interleaving
used to concatenate datasets, 50, *51-53*
used to interleave datasets, 53-54
Shapiro-Wilk test statistic, 107
SHORT option, CONTENTS procedure, 74
SKEWNESS option, MEANS procedure, *198*
SNK option, GLM procedure, *121*
SOLUTION option, GLM procedure, 144
Somer's D, 171
SORT procedure, 72-73
SPACE option, CHART procedure, 96
Spearman rank correlation, 79
Spearman's rho, 79
SPLIT option, PRINT procedure, 69
SQRT function, *189*
SSN format, *45*
STANDARD procedure, 195, 202-203
Standardized parameter estimates, 136
Statements, 9
STB option, REG procedure, 136, *137*
STD option
MEANS procedure, 102, *103*
in OUTPUT statement, *198*, 200
STANDARD procedure, 202, *203*
STDERR option, MEANS procedure, 102, *103*
in OUTPUT statement, *198*
Stem-and-leaf plots, 104, 107, 109
STEPDISC procedure, 83
Structural equation modeling, 82. *See also* CALIS procedure, SYSLIN procedure
Stuart's tau, 79
STUDENT diagnostic variable, REG procedure, 141
Student-Newman-Keuls multiple range test, *121*
Studentized residuals, 140
SUBMIT command, Display Manager, 267, *273*
Subsetting SAS datasets. *See* OUTPUT statement; WHERE statement
SUBSTR function, 225-226, 227
SUM function, *189*
SUM statement, PRINT procedure, 69, *70*
SUMMARY option, GLM procedure, 131
Sums of squares
Type I, 120, 127-128
Type III, 120, 127-128
Survival models, 81. *See also* LIFEREG procedure, ORTHOREG procedure
Syntax conventions, 2, 9
SYSLIN procedure, 82

T option, MEANS procedure, 102, *103*
t test, 79, 112
TABLES statement, FREQ procedure, 89, *90*, 156-157, *158*
TCSREG procedure, 81
TEST statement, REG procedure, 148, *149*
TIES option, RANK procedure, 204, *206*
Time-series data, 81
TITLE statement, 61, 62-63, 148, *236*
Tobit regression models, 81
TODAY function, *191*
TOP command, Display Manager, 268, *273*
TREE procedure, 83
TRIM function, 225, 228
TTEST procedure, 111, 112, 116-118
TUKEY option, GLM procedure, *121*
Tukey studentized range test, *121*
Two-group median test, *159*
Two-level dataset name, 28. *See also* LIBNAME statement

Unbalanced designs, 128
UNDO command, Display Manager, *273*
UNIFORM option, PRINT procedure, 69
Unit of observation, 17
UNIVARIATE procedure, 79, 98, 104-109, 111
compared to MEANS procedure, 102
default statistics, 106
Unplanned tests, 112
UP command, Display Manager, 268
UPCASE function, 225, 228-229
User-written formats. *See also* FORMAT procedure
compared to assignment statements, 255
User groups, 278
User services. *See* Help desks

VALUE statement, FORMAT procedure, 210-212, *213-217*
Van der Waerden scoring of ranks, *159*
VAR function, *189*
VAR option, MEANS procedure, 102
in OUTPUT statement, *198*, 199
VAR statement
MEANS procedure, 103, *196*, 197
NPAR1WAY procedure, 165, *166*
PRINT procedure, 65, 68, *70*